NAVAL
POSTGRADUATE
SCHOOL

MONTEREY, CALIFORNIA

THESIS

THE GREAT SPACE FORCE DEBATE:
A WAY FORWARD FROM THE PAST?

by

Jordan M. Johnson

December 2019

Thesis Advisor: James C. Moltz
Second Reader: Stephen H. Tackett

Approved for public release. Distribution is unlimited.

THIS PAGE INTENTIONALLY LEFT BLANK

REPORT DOCUMENTATION PAGE			*Form Approved OMB* *No. 0704-0188*
	Public reporting burden for this collection of information is estimated to average 1 hour per response, including the time for reviewing instruction, searching existing data sources, gathering and maintaining the data needed, and completing and reviewing the collection of information. Send comments regarding this burden estimate or any other aspect of this collection of information, including suggestions for reducing this burden, to Washington headquarters Services, Directorate for Information Operations and Reports, 1215 Jefferson Davis Highway, Suite 1204, Arlington, VA 22202-4302, and to the Office of Management and Budget, Paperwork Reduction Project (0704-0188) Washington, DC 20503.

1. AGENCY USE ONLY *(Leave blank)*	**2. REPORT DATE** December 2019	**3. REPORT TYPE AND DATES COVERED** Master's thesis	
4. TITLE AND SUBTITLE THE GREAT SPACE FORCE DEBATE: A WAY FORWARD FROM THE PAST?		**5. FUNDING NUMBERS**	
6. AUTHOR(S) Jordan M. Johnson			
7. PERFORMING ORGANIZATION NAME(S) AND ADDRESS(ES) Naval Postgraduate School Monterey, CA 93943-5000		**8. PERFORMING ORGANIZATION REPORT NUMBER**	
9. SPONSORING / MONITORING AGENCY NAME(S) AND ADDRESS(ES) N/A		**10. SPONSORING / MONITORING AGENCY REPORT NUMBER**	
11. SUPPLEMENTARY NOTES The views expressed in this thesis are those of the author and do not reflect the official policy or position of the Department of Defense or the U.S. Government.			
12a. DISTRIBUTION / AVAILABILITY STATEMENT Approved for public release. Distribution is unlimited.		**12b. DISTRIBUTION CODE** A	

13. ABSTRACT (maximum 200 words)

The concept of a "Space Force" has been debated in rooms at the Pentagon, on social media, and even as a potential series on Netflix, yet space operations have been an integral part of the U.S. armed forces for about 40 years. U.S. interest in space began as an urgent move to prove our nation's competitiveness during the Cold War. Today, the initiative to have a Space Force is rooted in the nationalism of President Donald Trump. The Space Policy Directive-4 signed by Trump is waiting for congressional approval, so a historical review of the role of space within the U.S. military is needed. This thesis analyzes the roles the Army, Air Force, and Navy have already played within the combined space effort. Through this historical analysis, this study offers a comparative examination of the space interests of each of the three branches of service, showing how each defines and views space, and addresses space organization within the services. The study concludes that the Air Force Space Command should remain under the U.S. Air Force and be renamed the Space Corps with both United States Space Command and the Space Corps being the central chain of command for all the services.

14. SUBJECT TERMS Air Force, Army, bureaucratic, Command and Control, Navy, policy, space, Space Force, Space Corps, space directive, United States, U.S. Space Command			**15. NUMBER OF PAGES** 107
			**16. PRICE CODE**
17. SECURITY CLASSIFICATION OF REPORT Unclassified	**18. SECURITY CLASSIFICATION OF THIS PAGE** Unclassified	**19. SECURITY CLASSIFICATION OF ABSTRACT** Unclassified	**20. LIMITATION OF ABSTRACT** UU

NSN 7540-01-280-5500

Standard Form 298 (Rev. 2-89)
Prescribed by ANSI Std. 239-18

THIS PAGE INTENTIONALLY LEFT BLANK

THE GREAT SPACE FORCE DEBATE:
A WAY FORWARD FROM THE PAST?

Jordan M. Johnson
Lieutenant, United States Navy
BA, Spelman College, 2013

Submitted in partial fulfillment of the
requirements for the degree of

MASTER OF SCIENCE IN SPACE SYSTEMS OPERATIONS

from the

NAVAL POSTGRADUATE SCHOOL
December 2019

Approved by: James C. Moltz
                Advisor

                Stephen H. Tackett
                Second Reader

                James H. Newman
                Chair, Department of Space Systems Academic Group

iii

THIS PAGE INTENTIONALLY LEFT BLANK

ABSTRACT

The concept of a "Space Force" has been debated in rooms at the Pentagon, on social media, and even as a potential series on Netflix, yet space operations have been an integral part of the U.S. armed forces for about 40 years. U.S. interest in space began as an urgent move to prove our nation's competitiveness during the Cold War. Today, the initiative to have a Space Force is rooted in the nationalism of President Donald Trump. The Space Policy Directive-4 signed by Trump is waiting for congressional approval, so a historical review of the role of space within the U.S. military is needed. This thesis analyzes the roles the Army, Air Force, and Navy have already played within the combined space effort. Through this historical analysis, this study offers a comparative examination of the space interests of each of the three branches of service, showing how each defines and views space, and addresses space organization within the services. The study concludes that the Air Force Space Command should remain under the U.S. Air Force and be renamed the Space Corps with both United States Space Command and the Space Corps being the central chain of command for all the services.

THIS PAGE INTENTIONALLY LEFT BLANK

TABLE OF CONTENTS

LIST OF FIGURES

THIS PAGE INTENTIONALLY LEFT BLANK

LIST OF ACRONYMS AND ABBREVIATIONS

ADCOM	Aerospace Defense Command
AFSPC	Air Force Space Command
AFSC	Air Force Systems Command
ARDC	Air Research and Development Command
ALB	Air Space Land Battle
ABM	Anti-Ballistic Missile
ASAT	Anti-satellite Weapons
ABMA	Army Ballistic Missile Agency
AOMC	Army Ordnance Missile Command
APL	Applied Physics Lab
ARGMA	Army Rocket and Guided Missile Agency
ASCC	Army Service Component Command
ARSTRAT/SMDC	Army Space and Missile Defense Command
USARSPACE	Army Space Command
ASI	Army Space Institute
ASPO	Army Space Program Office
ARSST	Army Space Support Team
CIA	Central Intelligence Agency
CFSCC	Combined Force Space Component Command
CSOC	Consolidated Space Operations Center
CSpOC	Combined Space Operations Center

CBRS	Concept Based Requirements System
DSCS	Defense Satellite Communications Systems
DoD	Department of Defense
DoDD	Department of Defense Directive
DCNO	Deputy of Chief Naval Operations
ELINT	Electronic Intelligence
FLTSATCOM	Fleet Satellite Communications
GPS	Global Positioning Satellites
GRAB	Galactic Radiation and Background High
HI-HO/HI-HOE	Altitude Program
HASC	House Armed Services Committee
ICBM	Intercontinental Ballistic Missile
IDC	Information Dominance Corps
ISR	Intelligence, Surveillance, and Reconnaissance
IWC	Information Warfare community International
IGY	Geophysical Year
JPL	Jet Propulsion Laboratory
JSpOC	Joint Space Operations Center
MOL	Manned Orbiting Laboratory
MIDAS	Missile Defense Alarm System
MUOS	Mobile User Objective System
NASA	National Aeronautics and Space Administration

NDAA	National Defense Authorization Act
NRO	National Reconnaissance Office
NSDC	National Space Defense Center
NAVWAR	Naval Information Warfare Systems Command
NPS	Naval Postgraduate School
NRL	Naval Research Lab
NAVSOC	Naval Satellite Operations Center
NSOC	Naval Space Operations Course
NAVSPACECOM	Naval Space Command
NAVASTROGRU	Navy Astronautics Group
PMW146	Navy Communications Satellite Program Office
NNSS	Navy Navigation Satellite System
NSC	Navy Space Cadre
NavSpaSur	Navy Space Surveillance System
NORAD	North American Aerospace Defense Command
OPCON	Operational Control
RAMMSO	Redstone Anti-Missile Missile System Office
SAMOS	Satellite and Missile Observation System
SecAF	Secretary of the Air Force
SASC	Senate Armed Services Committee
SAMSO	Space and Missile Systems Organization
SPAWAR	Space and Naval Warfare Systems Command

SWPC	Space Warfighter Preparation Course
SDI	Strategic Defense Initiative
TENCAP	Tactical Exploitation of National Capabilities
USAF	United States Air Force
AAF	United States Army Air Forces
USNORTHCOM	United States Northern Command
USSPACECOM	United States Space Command
USSTRATCOM	United States Strategic Command
USSF	United States Space Force
WDD	Western Development Division
WCDMA	Wideband Code Division Multiple Access
WGS	Wideband Global System

ACKNOWLEDGMENTS

My grace is sufficient for you, for my power is made perfect in weakness. Therefore I will boast all the more gladly of my weaknesses, so that the power of Christ may rest upon me. For the sake of Christ, then, I am content with weaknesses, insults, hardships, persecutions, and calamities. For when I am weak, then I am strong.

2 Corinthians 12:9-10

To my thesis advisor, Dr. Moltz, and second reader, Professor Tackett, thank you for your support and feedback throughout this process. My time here at NPS has been a roller coaster, full of ups and downs, and twists and turns. Thank you to all who have backed me during these times, especially CDR William Crane.

To my wonderful husband, Christian, and sweet daughter, Harper, thank you for your constant love. You are always everything I need and want in one. To the rest of my family, thank you for praying. God continues to show himself strong.

I dedicate this thesis to my son, the late Christian Rashad Johnson Jr., affectionately known as Duecey, my forever angel baby.

THIS PAGE INTENTIONALLY LEFT BLANK

I. INTRODUCTION

A. MAJOR RESEARCH QUESTION

Recently, the concept of a "Space Force" has been debated in rooms at the Pentagon, on social media, and even as a potential series on Netflix, yet space has already been an integral part of the U.S. armed forces for more than 40 years. Initially, interest in space within the U.S. began as an urgent move to show our national dominance. Pride in the nation was the driver of U.S. space directives in the early years. Today, similarly, the initiative to have a Space Force is rooted in the nationalism of President Donald Trump. He is an "America first" individual who is unapologetic about U.S. exceptionalism. His desire for a big, or even bigger, American presence on the world stage, cannot be missed in the abruptness of this call for a Space Force. Though it has been spoken of within military circles for years, it is very new to public ears, and hard for even the most knowledgeable to swallow. With the Space Policy Directive-4 signed by President Trump and waiting for Congress' approval, and the HASC (House Armed Services Committee) and SASC (Senate Armed Services Committee) having amendments to create a branch of the military for space under the Air Force, a historical review of the role of space within the U.S. Army, Air Force, and Navy was needed.[1] This thesis analyzes the roles each service has already developed within the combined space effort.[2]

This thesis analyzes the current debate regarding a Space Force by considering how it has been discussed in the past, as well as by comparing contributions and shortfalls of the three U.S. services in space. This study first discusses the history of space and important space assets in each branch of service. By providing a historical framework for the three services, this thesis shows how perspectives have altered with time and gives a more

---

[1] Sandra Erwin, "House Armed Services Markup to Bring Space Force Closer to Reality," *Space News*, last modified June 10, 2019, https://spacenews.com/house-armed-services-markup-to-bring-space-force-closer-to-reality/.

[2] Marcia Smith, "Text of Space Policy Directive-4 (SPD-4): Establishing a U.S. Space Force," Space Policy Online, February 19, 2019, (Presidential Memorandum. White House, 2019) https://spacepolicyonline.com/news/text-of-space-policy-directive-4-spd-4-establishing-a-u-s-space-force/.

accurate depiction of how the debate should proceed. Through this historical analysis, the study offers a comparative examination of the current space interests of each of the three branches of service, showing how each defines and views space, and discusses space organization within the services. The study then discusses the current Space Force debate and culminates by considering recommendations presented to improve the effectiveness of future policy from an operational standpoint.

B.   SIGNIFICANCE OF THE RESEARCH QUESTION

While internal discussion of space organization has been ongoing in military groups for many years, there is very little overt discussion about the current debate from military space professionals. Though there has been some to speak out, most of what is discussed openly is coming from the mouths of politicians and endowed meme influencers who lack critical knowledge about what the creation of a sixth branch in the military means for the warfighter strategically and operationally. The Space Directive is very vague in nature. It fails to outline major concerns, such as where previously established space assets from the respective services will lie once the Space Force is stood up. Former Defense Secretary Jim Mattis once said, "At a time when we are trying to integrate the Department's joint warfighting functions, I do not wish to add a separate service that would likely present a narrower and even parochial approach to space operations."[3] Though he subsequently altered his opinion on this matter to a supportive one, discussion of the negative side is still important. If the defense secretary once said it, along with a myriad of others, the statement holds some type of merit and should be discussed. Another question one would need to answer is what happens to the space cadres, space curricula, and space assets of respective services when the Space Corps is stood up? Will only members of the Space Corps be able to study space within the armed forces?

The significance of this thesis lies in understanding the nature of the space battlefield and the potential risk to military operations. This is important because space is a vital part of many missions in today's forces and without a proper understanding of how

---

[3] Helene Cooper, "A Space Force? The Idea May Have Merit, Some Say," *The New York Times*, last modified June 23, 2018, https://www.nytimes.com/2018/06/23/us/politics/trump-space-force-military.html.

to move forward on this issue, U.S. security will suffer. The last time the United States faced similar actions was during the transition to the current Air Force in the mid-1940s, when it broke off from the United States Army. It first commenced its initial separation as the Army Air Force Corps, then as the Army Air Force, which were both within the Army, until it made a final break away and became its own branch.[4] In this transition, analyst Mike Lorrey mentions major players of this change, such as General Billy Mitchell,[5] a significant subject matter expert and advocate of airpower. These trailblazers were participants in many of the debates with air power that now are relevant to space power. Though this transition did not happen as smoothly as it reads on paper, the cohesiveness between military and civilian sectors is what created the successful Air Force we see today. Since the voices of military space experts are not being widely consulted in the current debate, this thesis will help their voices to be heard once again.

With direction from the White House to establish the sixth and smallest branch of the Department of Defense, the Space Force, has brought up many multifaceted questions. On February 19, 2019, President Trump signed the Space Policy Directive-4 (SPD-4)[6] and is currently awaiting congressional approval.

The "U.S. Space Command Talking Points and RTQ"[7] (response to query), states that

> Establishing a combatant command focused on space is complementary to establishing a new branch of the Armed Forces for space. Once established, a U.S. Space Force would organize, train, and equip space forces, and U.S. Space Command would employ these forces to compete, deter, and win.[8]

---

[4]Christopher McCune, "A Brief History of the U.S. Air Force," Air Force Space Command, last modified September 12, 2016, https://www.afspc.af.mil/News/Article-Display/Article/942428/a-brief-history-of-the-us-air-force/.

[5]Mike Lorrey, "Op-ed: The Legal Mandate for a U.S. Space Force," *Space News*, last modified October 26, 2018, https://spacenews.com/op-ed-the-legal-mandate-for-a-u-s-space-force/.

[6]Smith.

[7] Department of the Air Force, "U.S. Space Command Talking Points and RTQ," STRATCOM, 12 April 2019.

[8] Department of the Air Force.

The standup of U.S. Space Command would require pulling space experts from all branches of the military to support, and is the first step of many toward a proposed U.S. Space Force. This reestablishment is meant to accelerate efforts to incorporate more space capabilities operationally. The stand-up of U.S. Space Command as the 11th combatant command should help speed up space operations, as it will be the center of control for it.[9]

Though President Trump sees this as forward momentum, this thesis will show why a dedicated Space Force may or may not be needed, as the presence of space within the forces has been longstanding. Also, though just entering the public eye, the Space Force debate is not new. In the 2001 Space Commission report, two proposed alternatives were debated for the way moving forward for space study within the military. The two proposed statements, an independent service dedicated to space completely apart from the U.S. Air Force, or a Space Corps within the U.S. Air Force were intensely deliberated. This ended in neither occurring, as the commissioners agreed to move forward with the U.S. Air Force being the best service to cultivate space possibilities for the foreseeable future.

This sounds familiar as it is what lawmakers are currently discussing almost two decades later now that President Trump has made the announcement to call for the creation of the Space Force. There are many views surrounding this conversation and the three inferences within the Space Commission report are the same circling around today and will be discussed as hypotheses in this thesis.

In *Mastering the High Ground*, military space experts such as Lieutenant Colonel Bruce DeBlois argue the need for an independent Space Force due to the different requirements that air power and space power have from a doctrinal level.[10] U.S. Air Force Col. (ret.) Whittington also agrees that the U.S. Air Force has applied traditional air power terminology to space.[11] For example, he says, "offensive counter air becomes offensive

---

[9] Everett C. Dolman, "Space Force Déjà Vu," *Strategic Studies Quarterly 13*, no. 2 (2019): 16, https://www.airuniversity.af.edu/Portals/10/SSQ/documents/Volume-13_Issue-2/Dolman.pdf.

[10] Benjamin Lambeth, *Mastering the Ultimate High Ground* (Santa Monica, CA: RAND,2003),69,https://www.rand.org/content/dam/rand/pubs/monograph_reports/2005/MR1649.pdf.

[11] Michael C. Whittington, "A Separate Space Force, an 80-year-old Argument," Maxwell Paper 20, no. 1 (May 2000): 9, Air War College.

counter space." He disagrees that this diction is the way forward because believes that "space power theory and doctrine cannot be built on an aviation foundation."[12] Other arguments have stemmed from the claimed need for military space to be centralized, meaning by having a Space Force, such that redundancy will be decreased within other forces and space can be single focused for decision-making and funding.[13] U.S. Air Force Col. (ret.) Terry Virts (a former astronaut) argues that space has been its own domain for years and that if the U.S. fails to take extreme measures, such as creating a completely separate Space Force, the U.S. will fall behind in being the "world's leader in space,"[14]with China and Russia emerging as definite threats. His recommendations include consolidating certain missions, such as the launching and controlling satellites, as well as all cyber forces, within the Space Force.[15] The main point of this effort would be to "free space from the inhibiting bonds of traditional air power thinking."[16]

On August 9, 2018, Vice President Mike Pence outlined President Trump's proposed Space Force and the threats posed by China and Russia.[17] The arguments in support of an independent force focus on the importance of staying ahead of these adversary countries in space centralizing U.S. military efforts in a single organization.

By contrast, Major General James B. Armor recommends a Space Corps to address some of the same concerns. He makes the point that just as the U.S. Air Force needed to separate doctrine from the Army in order to gain airpower and reduce the Army's influence, space power needs its own doctrine.[18] With a Space Corps, this would allow the U.S. Air

---

[12] Whittington, 9.

[13] Lambeth, 70.

[14] Terry Virts, "I Was an Astronaut. We Need a Space Force," *The Washington Post*, August 23, 2018, https://www.washingtonpost.com/opinions/i-was-an-astronaut-we-need-a-space-force/2018/08/23/637667e6-a6fb-11e8-b76b-d513a40042f6_story.html?utm_term=.a6d767e3e16e.

[15] Virts.

[16] Lambeth, 68.

[17] Mike Pence, "VP Mike Pence Speech on SPACE FORCE Pentagon August 9, 2018," August 9, 2018, produced by Alberto Lopez, YouTube, 25:47:00, https://www.youtube.com/watch?v=X6maYkfZ514

[18] James B. Armor, Jr., "Viewpoint: It is Time to Create a United States Air Force Space Corps," *Astropolitics* 5, no. 3, (September-December 2007), 276, doi: 10.1080/14777620701580851:280-281.

Force to utilize the support, logistically speaking, that they already have stood up and not have to reinvent the wheel completely.[19] This is easier to visualize from an administrative standpoint to some because, though the Space Corps would be the Department of Defense's only space establishment, it would still fall under the Air Force, making it easier to control organizationally and, arguably, cheaper. Over the years, Congressman Mike Rogers has been a heavy influencer of the need for a Space Corps. Back in 2013, he emphasized that Air Force spending was declining in regard to space but increasing for non-space agendas.[20] He has been arguing since then that the Air Force has other priorities that seem to take precedence over space and to change this narrative, a Space Corps is needed because of threats we are now facing. In his 2017 space symposium speech, he stresses that "potential adversaries are developing weapons to take out our own space systems in a conflict."[21] He does not believe that the Air Force makes space a priority and that creating a separate service would "produce clearer lines of responsibility and accountability."[22] From a bureaucratic standpoint, he does argue that is will be unsettling in the short term,[23] but worth it later to see us above our adversaries with space operations.

Many analysts and officials believe the United States needs neither a Space Force nor a Space Corps. Senator (Dem., Mass.) Elizabeth Warren, who is a candidate for the 2020 Democratic presidential nomination, has publicly expressed skepticism about the Space Force and argues that it will only waste money.[24] This raises the question that if President Trump leaves the White House before congressional approval or significant action to stand up the Space Force, the drive to create it may dissipate as quickly as it

---

[19] Lambeth, 69.

[20] Wilson Brissett, "The Space Corps Question," *Air Force Magazine*, last modified October 2017. http://www.airforcemag.com/MagazineArchive/Pages/2017/October%202017/The-Space-Corps-Question.aspx.

[21] Mike Rogers, "Remarks of Congressman Mike Rogers," *Strategic Studies Quarterly 11*, no. 2 (2017): 4, https://www.airuniversity.af.edu/Portals/10/SSQ/documents/Volume-11_Issue-2/Rogers.pdf

[22] Brissett.

[23] Brissett.

[24] Robert Burns, "Senators on Trump Space Force Plan: Not so Fast," Associate Press, April 11, 2019, https://www.apnews.com/5a6e1a08acb745a69976e3110b4789b4.

began. Former NASA director Sean O'Keefe calls the proposal an expense that is not needed.[25] This is one of the main reasons some are opposed to both the Space Force and Corps, because of the financial burden it is expected to have. With an initial estimated cost of $13 billion[26] over the next five years (an estimate released from the Office of the Secretary of the Air Force), the argument is that space resources already available in the respective services would be better spent in other ways and that, in the end, space is not yet a location for force application, but is a supporter, or a facilitator, to other operations. Other critics discuss the projected small size of the Space Force, which would likely total only some 15,000 by 2025. By any comparison, this small size, does not match up with a service that should receive separate funding.[27]

C.   POTENTIAL EXPLANATIONS AND HYPOTHESES

This thesis is centered on three hypotheses through the historical analysis. First, that there should be an independent space service, but after years as a corps under the U.S. Air Force. This is the current direction the Space Directive-4 is heading, if approved by Congress. Just as airpower was separated from the U.S. Army back in 1947, there are arguments that space power needs to be separated from the Air Force.

The first hypothesis discussed—in response to the Trump administration's recent proposal--is that there should by an independent space service, and that it should immediately be separated from the U.S. Air Force. It is said that threats coming from adversary countries will only get worse and the U.S. must guarantee operationally the ability to protect itself and assets in space. Just like air, land, and sea are their own domains, arguments that space is as well and should be treated as such will be discussed.

A second hypothesis, drawing on the work Gen. Armor and the congressional legislation initially pushed by Congressman Rogers, is that there be an initial Space Corps under the Air Force, just like the Army Air Corps was under the United States Army. This

---

[25] Dolman.

[26] Burns.

[27] Dolman.

7

would make it easier to integrate space operations with air, land, and sea, but without requiring space to break out into its own branch.

The third hypothesis discoursed in this thesis is that the Space Directive-4 should be rejected, and the services should remain on the current course, or adopt other reforms. Standing it up alone requires great effort, including funding and a completely separate hierarchy, which takes manpower.

Questions such as where the space cadre in current services will go, how efficient adding a branch will be, and how it will look from a bureaucratic standpoint are answered in this thesis. Clearly, separating the pros and cons of these different options will help decision makers construct the best choice moving forward.

D.   RESEARCH DESIGN

This thesis is a comparative analysis of the U.S. Army, Navy, and the Air Force and their roles in space. This allows insight into how the three largest branches of the military have viewed and treated space over the years. The goal of these chapters is to analyze comparatively the contributions to space of the respective services. The thesis then introduces the current space debate. The thesis culminates with addressing the hypotheses and questions stated above and then discusses possible policy implications. The methodological approach for this thesis is a set of comparative historical cases. Starting with the history of the Army in space is appropriate, since it is the oldest branch of the military, and ending with research in regard to the current space debate. This approach leaves out the Marine Corps and the Coast Guard because, statistically speaking, they are too small of a space presence to be included in this thesis, both in terms of manpower and budget.

E.   THESIS OVERVIEW

After the introduction, this thesis begins with the Army chapter. This chapter discusses the Army's history in relation to space and the different policies for space overtime. The third chapter is about the Navy and its evolving space understanding operationally. This chapter outlines the Navy's relationship with space as it ties back to the Cold War era and beyond. The fourth chapter is about the Air Force and its affiliation with space. It also discusses moving forward how space has played different roles in the growth of the Air Force. All of the Armed Forces chapters are in chronological order. The fifth chapter discusses the current space debate by referencing the Space Directive-4; going through it and commenting how this new change affects the different forces both directly and indirectly. It also discusses the current and potential future Command and Control (C2) organizational structures. It then deliberates hypotheses and the questions asked within the significance of the research question section. It also discusses the U.S. Space Command's and Space Development Agency's (SDA) inevitable bond with the Space Force. Policy implications and conclusions for future potential actions regarding the Space Force/Corps are given to suggest further questions for study of this topic.

THIS PAGE INTENTIONALLY LEFT BLANK

II. SPACE HISTORY AND ADVANCEMENTS WITHIN THE U.S. ARMY

A. INTRODUCTION

This chapter evaluates the history of the U.S. Army's contribution in space. Though the end of World War II introduced new threats, such as nuclear weapons, the U.S. Army's interest in space came from the desire for superiority against the Soviet Union in rocket technology and strategic defense, while becoming more knowledgeable through the development of sensor technology to provide intelligence from space assets. This chapter begins with the U.S. Army and Army Air Corp's early rocket work with the prominent Dr. Werner Van Braun, and the U.S. expedited interest in obtaining the Nazi's arsenal of rockets and other military research to safeguard this desired information from the Soviet Union. This chapter explains why the Army had an operational pause in space activity, and why it was able to renew its space pursuit later and continuing to the present. This chapter culminates with the Army's current role in space, discussing ongoing space research the U.S. Army is involved in, and also what specific space assets the Army operates. This chapter will help to assess space influence and assets within the Army to relate to the debate over a Space Force and to provide policy recommendations.

B. THE ARMY'S CLAIM TO SPACE

Toward the end of WWII, as Germany buckled in defeat, the U.S. was fervent to capture key German technological research development.[28] Germany unknowingly set the U.S. up for success in this quest when in March of 1945, the Osenberg List,[29] a verified document of scientists and technical personnel in Germany, was discovered in an

---

[28] James C. Moltz, *The Politics of Space Security: Strategic Restraint and the Pursuit of National Interests*, Second Edition (Stanford, CA: Stanford University Press, 2011), 85.

[29] "World War II: Operation Paperclip," Jewish Virtual Library, accessed September 16, 2019, https://www.defense.gov/About/. https://www.jewishvirtuallibrary.org/operation-paperclip.

improperly flushed toilet at Bonn University by a Polish laboratory technician.[30] This vital information fell into the hands of U.S. Intelligence, and it helped Major Robert Staver, who, at the time, was the chief of the Jet Propulsion Section in the Research and Intelligence Branch of the U.S. Army Ordnance,[31] and also known as the father of the U.S. covert program Operation Overcast/Paperclip.[32] In this program, headed by the Joint Intelligence Objectives Agency (JIOA)[33] approximately 1,000[34] German scientists and engineers along with their families were granted refuge in the U.S. Without the success of previous secret operations such as Special Mission V-2 and Operation LUSTY[35] the U.S. would not have been able to carry out Operation Overcast/Paperclip. In November of 1945, Operation Overcast was officially renamed Operation Paperclip by the U.S. Army because, when the Army officers came across a possible applicant amongst these rocketry experts, they would annotate the file by attaching a specific colored paperclip on it.[36]

Though many of the scientists were Nazi-party members, and U.S. rules prohibited the recruitment of any Nazis, the Secretary of Commerce at the time persuaded President Harry Truman to approve them despite the obvious misrepresentation and concealment of

---

[30] Anne Jacobsen, Operation Paperclip: The Secret Intelligence Program that Brought Nazi scientists to America (New York: Little, Brown and Company, 2014), chap.3.

[31] "Wernher Von Braun," New World Encyclopedia, last modified October 19, 2016, https://www.newworldencyclopedia.org/entry/Wernher_von_Braun.

[32] Allison Eldridge-Nelson, "Veil of Protection: Operation Paperclip and the Contrasting Fates of Wernher Von Braun and Arthur Rudolph," (master's thesis, Graduate College of Bowling Green State University, 2017), 40, https://etd.ohiolink.edu/!etd.send_file?accession=bgsu1510914308951993&disposition=inline.

[33] Eldridge-Nelson, 37.

[34] Katie Serena, "How Nazi Scientist Wernher Von Braun Sent the U.S. To The Moon," ATI, last modified December 18, 2017, https://allthatsinteresting.com/wernher-von-braun.

[35] Jerry V. Drew II, "First in Space: The Army's Role in U.S. Space Efforts, 1938–1958," (space master's thesis, Naval Postgraduate School, 2017), 25, https://pdfs.semanticscholar.org/4f39/4d47720551819978d26fc185ccb1a8cba553.pdf.

[36] Clarence G. Lasby, *Project Paperclip: German Scientists and the Cold War* (New York, 1971), 155.

the agency to make these scientists check out as "legitimate."[37] Among those scientists was Dr. Werner Von Braun, the former technical director at the Peenemunde Army Research Center in Germany,[38] and the man credited with the development of the V-2 rocket, known as the "vengeance weapon."[39] Though Von Braun later brought great success to the U.S. in space, it is necessary to note that he undoubtedly was aware of the happenings in the concentration camps at the time, and the process of handpicking people from the camps to work inside the Peenemunde center building his rockets, where thousands died from the harsh conditions.[40] Though this history is appalling, Major Staver believed that "future scientific importance outweighs their present war guilt,"[41] as a report to the U.S. State Department said. The V-2 rocket was the first man- made object to travel into space rising to an altitude of 118 miles,[42] and also the rocket that resulted in the death of over 5,400 civilian and military personnel in neighboring cities during WWII by German hands.[43] Eager to be absolved of all guilt and start a new life in the U.S., in late 1945, 127 German scientists, including Von Braun, arrived at Fort Bliss, Texas, to work on perfecting the V-2 rocket alongside of Project Hermes,[44] a two-stage missile seeking to improve the V-2. The V-2 saw multiple test flights during the time, with the first success occurring on May 10, 1946, in which the rocket hit an altitude of 70 miles at the White Sands facility in

---

[37] Aaron Leonard, "The Horrible Secrets of Operation Paperclip: An Interview with Annie Jacobsen about Her Stunning Account," History News Networks, last modified 2019, https://historynewsnetwork.org/article/155194.

[38] "Peenemünde Army Research Center," Atlas Obscura, accessed September 16, 2019, https://www.atlasobscura.com/places/peenemunde-army-research-center-1.

[39] Moltz, 70.

[40] Jacobsen, Chap 21.

[41] Victor Sebestyen, *1946: The Making of the Modern World* (New York: Knopf Doubleday Publishing Group, Nov 29, 2016), 244.

[42] Suzanne Deffree, "German Rocket Is 1st to Reach Space," October 3, 1942,"EDN Network, last modified October 3, 2018, https://www.edn.com/electronics-blogs/edn-moments/4397678/German-rocket-is-1st-to-reach-space--October-3--1942.

[43] Moltz, 70.

[44] Drew, 34.

New Mexico.[45] In November of 1948,[46] the Germans relocated to Redstone Arsenal, next to Huntsville, Alabama, which would soon be known as "Rocket City USA."[47] Von Braun and his colleagues soon got to work creating V-2 byproducts, which were still tested at White Sands Missile Range. One of these by-products was a ballistic missile based off the Hermes C project,[48] one of the Redstone studies. In order to speed up development in hopes of making it a field weapon in the Korean War, a rocket booster was used and, in 1952, this missile was officially renamed Redstone. After 11 test flights, in March of 1956 the Redstone booster reached space and the Army's first Redstone battalion was formed.[49]

The Army Ballistic Missile Agency (ABMA) was established at the Redstone Arsenal on February 1, 1956, and was commanded by General John B. Medaris, and the technical director was Dr. Werner Von Braun.[50] Though the U.S. Army had already reached space with the Redstone booster in 1956, this team was crucial in helping the U.S. earn respect in space after the U.S. Navy's unsuccessful Vanguard launch attempt in December of 1957. With help from the Jet Propulsion Laboratory (JPL),[51] they successfully launched *Explorer 1* on January 31, 1958, using the Juno I (a modified Redstone missile).[52] It was confirmed that *Explorer 1* had completed its first orbit around the Earth roughly two hours after the launch.[53] After the launch, Von Braun stated, "We

---

[45]Peter L. Eidenbach, "A Brief History of White Sands Proving Ground: 1941–1965,"accessed September 16, 2019, 8, https://web.archive.org/web/20101229102456/http://nmsua.edu/tiopete/files/2008/12/wspgcoldbook.pdf.

[46] Drew, 49.

[47]"Huntsville, Alabama "Rocket City, USA" A Spirit of Exploration," March 23, 2019, Unravel Travel TV, 2:17:00, https://www.youtube.com/watch?v=KsKsU1VeexM.

[48] Andrew J. LePage, "Old Reliable: The Story of the Redstone," The Space Settlement Enterprise, last modified May 2, 2011, http://www.thespacereview.com/article/1836/1.

[49] LePage.

[50] Jason B. Cutshaw, "Celebrating the Army's Explorer 1 Legacy," U.S. Army, last modified January 30, 2018, https://www.army.mil/article/199846/celebrating_the_armys_explorer_1_legacy.

[51] Cutshaw.

[52] LePage.

[53] Cutshaw.

have now established our foothold in space, we will never give it up again."[54] Dr. James A. Van Allen, a space scientist at the University of Iowa is known for developing the scientific instruments on *Explorer 1*, which was the first satellite to gather information on Earth's radiation belts, later named the Van Allen radiation belts after him.[55]

The DoD's 1957 decision to transfer initial space-launch responsibilities[56] to the Army came with the formation of the Army Redstone Anti-Missile Missile System Office (RAMMSO)[57] on October 3, 1957, ironically one day before Sputnik. This office was responsible for the management of different missile defense systems.[58] This office also led to the formation of Nike-Zeus, an anti-missile system designed to demolish Soviet ICBMs.[59] Though there was a trend away from Anti-satellites (ASATs), the Secretary of Defense under the Kennedy administration, Robert S. McNamara pushed for it. The initial adjustment happened in May of 1962 in which the U.S. Army 505 program altered the Nike Zeus to carry nuclear warheads that could destroy satellites in Low Earth Orbit (LEO).[60] However, just seven months after RAMMSO's establishment, it underwent an organizational restructure, as the Army decided it wanted all Redstone missile programs to

---

[54] Cutshaw.

[55] Cutshaw.

[56] Joshua Boehm, "A History of United States National Security Space Management and Organization," Commission paper, Commission for Assessment of United States National Security Space Management and Organization, 2001, 23,https://fas.org/spp/eprint/article03.html#ft78.

[57] SMDC/ARSTRAT Public Affairs, "SMDC Celebrates 60 Years of Space and Missile Defense Excellence," U.S. Army, last modified September 21, 2017, https://www.army.mil/article/194216/smdc_celebrates_60_years_of_space_and_missile_defense_excellenc e.

[58] SMDC/ARSTRAT Public Affairs.

[59] James Walker, Lewis Bernstein, and Sharon Long, *Seize the High Ground: The Army in Space and Missile Defense* (U.S. Army Space and Missile defense Command, 2003), 27, https://history.army.mil/html/books/070/70-88-1/cmhPub_70-88-1.pdf.

[60] Bruce M. DeBlois, "Space Sanctuary: A Viable National Strategy," *Air and Space Power Journal,* no. 1 (1998): 45, https://cle.nps.edu/access/content/group/19ebe878-76c5-4974-b810- c080ee8102f6/Course%20Documents/Class%2012/DeBlois1998.pdf.

be under one commander.[61] This restructure was very interesting because it was merged in with the Army Rocket and Guided Missile Agency (ARGMA), whose commander, Maj. Gen. John G. Shinkle[62] was the very man who petitioned and promoted the commissioning of RAMMSO. Though RAMMSO's life was very brief, significant programs developed within it, such as Nike-Zeus, were transferred on.[63] With the U.S. Army in a constant state of restructure and focusing on the importance of aggregating the study of missiles,[64] ABMA, ARGMA, and the JPL were soon subordinate commands under the U.S. Army Ordnance Missile Command (AOMC), formally stood up on August 1, 1958.[65]

1. The Army's Space Ambitions dissolve into NASA

On October 21, 1958, Vice President Richard Nixon convinced President Dwight D. Eisenhower that a civilian agency was needed to direct space related activities. When Congress approved this shift, the AOMC lost all space related tasks to the newly established National Aeronautics and Space Administration (NASA).[66] Even though the bulk of its prior space missions were transferred to NASA, ABMA continued to work on space related jobs. One was the Project Horizon in June of 1959, which proposed a manned military base on the moon.[67] Congruently, General Medaris submitted a 15-year space program plan based on the development of heavy boosters made by the ABMA. In *The Heavens and the Earth*, author Walter McDougall states that the report forecasted "lunar

---

[61]Mark Hubbs, "Mission Begins on Redstone Road," *The Eagle*, October 2007, 12, https://www.smdc.army.mil/Portals/38/Documents/Publications/History/Eagle%20articles/SMDCMissionBeginsonRedstoneRoad.pdf?ver=2019-01-11-144934-020.

[62]Hubbs, 12.

[63]Hubbs, 12.

[64]"The ABMA/AOMC Era, 1956–62,"AMCOM, accessed May 23, 2019, https://history.redstone.army.mil/ihist-1956-intro.html.

[65]U.S. Army, "AMCOM History," U.S. Army, last modified December 4, 2014, https://www.army.mil/article/139391/amcom_history.

[66]U.S. Army.

[67] Arthur C. Trudeau, *Proposal to Establish a Lunar Outpost* (U.S. Army: Chief of Research and Development, 1959), https://history.army.mil/faq/horizon/Horizon_V1.pdf.

reconnaissance and two-man satellites by 1962, manned lunar circumnavigation by 1963, and a fifty-man moon base by 1971."[68] Project Horizon was rejected, however, and in 1961, ABMA and ARGMA were completely eliminated, as they were viewed as not needed anymore due to NASA inheriting all space work from AOMC, including the Saturn booster plans within Project Horizon.[69]

C.   THE VIETNAM WAR: ARMY'S PAUSE ON SPACE DEVELOPMENTS

With the Vietnam War being in full throttle, the Army's research endeavors were refocused away from space and back toward surface-to-surface missiles.[70] Though the Army's main priority was developing weaponry that the soldiers could use tactically during the war, the Army still made contributions to developing a global communications linkage for the DoD.[71] This connected the Army to contributing to the first geosynchronous satellite, *SYNCOM*.[72] Space was still in the fight indirectly, and that is why in 1964 the Army established the Strategic Communications Command, to answer the demand for satellite communications.[73] The war allowed satellite communications to be used operationally from different areas of operations.[74] By 1973, the Army had almost completely withdrawn from Vietnam.[75] Though this was only three years after the

---

[68] William A. McDougall, *The Heavens and the Earth: A Political History of the Space Age*, Baltimore, MD: The John Hopkins University Press, 1985, http://asan.space/wp-content/uploads/2018/05/Walter-A.-McDougall-The-Heavens-and-the-Earth_-A-Political-History-of-the-Space-Age-1997-Johns-Hopkins-University-Press.pdf.

[69] Trudeau.

[70] Boehm, 24.

[71] Walker, Bernstein, and Long, 48.

[72] Walker, Bernstein, and Long, 48.

[73] Boehm, 24.

[74] David N. Spires and Rick W. Sturdevant, *Beyond the Ionosphere: Fifty Years of Satellite Communication* (Washington, DC.: The NASA Series, 1997), chap. 7, https://history.nasa.gov/SP-4217/ch7.htm.

[75] Walker, Bernstein, and Long, 48.

Department of Defense Directive (DoDD) 5160.32[76] was revised to allow the other forces (along with the Air Force) to develop space assets that met their specialized warfighting needs,[77] the Army was less concerned about space and more fixated on the improvements for ballistic missile defense.[78] Though the Army Space Program Office (ASPO) was established in 1973,[79] not much came from it except plans to execute the Tactical Exploitation of National Capabilities (TENCAP) program,[80] which was not officially mainstream until Congress's decision to model all of the U.S. Armed services' TENCAP programs off of the original Army plans in 1977.[81]

D. THE ARMY SPACE REBIRTH AND ADVANCEMENTS

The TENCAP program was made up of multiple programs contributed by each of the services with the goal of influencing and integrating national space resources with intelligence data provided from different space systems.[82] By 1982, the development of a Concept-Based Requirements System (CBRS) for space acquisitions and the Air Space Land Battle (ALB) Doctrine helped the Army's further pursuit of space interests.[83] On March 23, 1983, President Ronald Reagan announced the Strategic Defense Initiative (SDI), the program known as "Star Wars."[84] This catapulted the Army's developments of

---

[76] Department of Defense, Reconnaissance, Mapping and Geodetic Programs Reference, DoD Directive 5160. 32 (Washington, DC: Department of Defense, 1961), https://www.nro.gov/Portals/65/documents/foia/declass/WS117L_Records/215.PDF.

[77] Boehm, 24.

[78] Walker, Bernstein, and Long, 48.

[79] John Pike, "Tactical Exploitation of National Capabilities (TENCAP)," Global Security, last modified July 28, 2011, https://www.globalsecurity.org/intell/systems/tencap.htm.

[80] Pike.

[81] Pike.

[82] Boehm, 24.

[83] Eddie Mitchell, *Apogee, Perigee, and Recovery: Chronology of Army Exploitation of Space*, (Santa Monica, Ca:), 75, https://apps.dtic.mil/dtic/tr/fulltext/u2/a254602.pdf.

[84] "Strategic Defense Initiative (SDI)," Atomic Heritage Foundation, July 18, 2018, https://www.atomicheritage.org/history/strategic-defense-initiative-sdi.

Anti-Ballistic Missile (ABM) technology and capabilities designed to shoot missiles from space.[85] SDI also opened the Army's eyes to the need for enhancing and using space assets to improve its effectiveness. The 1983 Army Science Board study, "Army Utilization of Space Assets," determined that space was not being taken advantage of fully by the Army. Nevertheless, by December 1985, space interest within the Army had risen to a point that an Army space initiatives study recommended formation of an Army Space Command as a subset of U.S. Space Command, and making the Office of Deputy Chief of Staff for operations and plans the senior Army staff space advocate.[86] This initiative foreshadowed the Army's space cadre official development years later, as it expressed a need to "train soldiers about space systems and create an additional specialty indicator to trace personnel with experience, education, and training in space systems."[87]

The study's recommended actions, many of them implemented between the years of 1986 and 1987,[88] included the founding of the Army Space Institute and Training and Doctrine Command,[89] both relating to the education and use of space in relation to the Army. The Army Space Agency was also formed out of this study in 1986, although in 1988 its title was changed to the Army Space Command (USARSPACE)[90] with a goal of coordinating Army space-related endeavors with USSPACECOM.[91] By 1992, there were so many space programs and entities within the Army that the Army Science Board came back and recommended a consolidation of its space assets. Thus, in 1997, the Army Space and Missile Defense Command (ARSTRAT/SMDC)[92] was formed and became the

---

[85] Mitchell, X.

[86] Walker, Bernstein, and Long, 7.

[87] Walker, Bernstein, and Long, 7.

[88] Boehm, 23.

[89] Mitchell, 77.

[90] Mitchell, 77.

[91] Boehm, 24.

[92] SMDC/ARSTRAT Public Affairs.

primary Army Service component command (ASCC) to give space updates to United States Strategic Command (USSTRATCOM), which had been formed in 1992.

1.   Formation of Space Cadre and Current Space Research and Assets

Over the years, Congress became increasingly concerned about how space assets were managed, as well as the organizational structures that oversaw it. Thus, in 1999, Congress created the Rumsfeld Commission to examine these issues.[93] This commission's report, officially issued in January 2001, offered both an evaluation on different inadequacies within the DoD military in regard to space, as well as suggestions to organize forces in order to better support.[94] In this section, commissioners connected three main issues, one of which was the need for "cadre-building,"[95] that resulted in tasking the Army to develop a space cadre.[96] Though the Rumsfeld Commission report was released in 2001, the Army had recognized a need for a dedicated space cadre well before.[97] The Army created the space operations functional area (FA-40) and comprises 389 officers on active-duty, in the Army National Guard, and in the U.S. Army Reserve.[98] The term "Functional Area" within the Army represents soldiers that concentrate for their entire career in that designated specialty.[99] Though the Army has this designated functional area for space, U.S. Army personnel outside of the FA-40 community support space assets as well. For example, the Army has about 2,500 personnel active-duty officers and enlisted, reserves,

---

[93]Marcia S. Smith, *Military Space Activities: Highlights of the Rumsfeld Commission Report and Key Organization and Management Issues*, CRS Report No. RS20824 (Washington, DC: Congressional Research Service, 2001), *https*://www.everycrsreport.com/files/20010221_RS20824_f0489c442dba0986a6d7e3ca843938b52559bc9 d.pdf

[94] Lambeth, 63.

[95] Lambeth, 63.

[96] Lambeth, 63.

[97]Kevin Campbell, "The Army's Space Cadre," *High Frontier Journal* vol. 4, no. 1 (November 2007): 3, https://documents.theblackvault.com/documents/space/AFD-071119-017.pdf.

[98] "FA40 Roster," ASPDO, August 21, 2019.

[99] Campbell, 3.

and civilians involved in space operations at ARSTRAT/SMDC.[100] It was also just recently announced that U.S. Army Lieutenant General James Dickinson has been nominated to serve as the Deputy Commander of the newly re-established U.S. Space Command (USSPACECOM),[101] a command that originally existed from 1985 until 2002, but was combined with USSTRATCOM when focuses shifted to homeland security after the 9/11 attacks.[102] The re-established 2019 USSPACECOM is a step to highlight the importance of the space domain as well as what it contributes to all of the forces.[103]

Today, the U.S. Army remains active in providing space support to the warfighter. The Army is interested in making sure space functions are taught throughout the forces so that a preponderance of personnel can be knowledgeable users. The Army is constantly growing its space cadre, as more personnel find interest in space. It is also expedient that the FA-40 designator has its own promotion metric,[104] meaning one can take an exclusive space track without worrying about career progression.

Though the support via personnel is impressive, the Army's involvement in satellites currently in orbit is also notable. In 2010, after 50 years of silence in this area, the U.S. Army flew the SMDC-ONE CubeSat.[105] Though this mission lasted only 35 days, it paved the way by demonstrating that CubeSats were dependable. Since this mission, the

---

[100] Sandra Erwin, "Majority of Army's Space Soldiers will not Transfer to the Space Force," Space News, last modified May 15, 2019, https://spacenews.com/majority-of-armys-space-soldiers-will-not-transfer-to-the-space-force/.

[101] Jen Judson, Jeff Martin, and Joe Gould, "Here's Who Will Be the U.S. Space Command Deputy," Defense News, last modified August 6, 2019, https://www.defensenews.com/digital-show-dailies/smd/2019/08/06/heres-who-will-be-the-us-space-command-deputy/.

[102] Tracy Cozzens, "U.S. Space Command Re-Established as 11th Unified Combatant Command," GPS World, last modified August 29, 2019, https://www.gpsworld.com/u-s-space-command-re-established-as-11th-unified-combatant-command/.

[103] Cozzens, "U.S. Space Command."

[104] Department of the Army, Commissioned Officer Professional Development and Career Management, Pamphlet 600–3 (Department of the Army, December 3, 2014),10, https://www.army.mil/e2/c/downloads/376665.pdf.

[105] Billy E. Johnson and Martin F. Lindsey, U.S. Army Small Space Update, SSC16-III-06 (SMDC-PA No. 6085, June 3, 2016), https://digitalcommons.usu.edu/cgi/viewcontent.cgi?article=3349&context=smallsat.

U.S. Army has launched three SMDC Nanosatellite Program-3 (SNaP) in October of 2015 and are all currently still on orbit.[106]

There are also constant advancements being developed at ARSTRAT/SMDC that are maintained within the future warfare center.[107] With the ever- increasing need for reliable transmissions, another technological advancement SMDC is experimenting with is the U.S. Army's Kestrel Eye program.[108] This microsatellite technology is a limited duration LEO satellite operated at the brigade level within the U.S. Army. This microsatellite gives electro-optical images to warfighters within the area by way of a direct downlink. This is expedient because it satisfies the situational awareness needs of the warfighter via tactical imagery instantly. In October of 2017, the Kestrel Eye was deployed from the International Space Station and is up and running and in a testing period.[109] Many national systems are not as responsive for timely decisions on the battlefield, so the Kestrel Eye program, now known as Kestrel Eye Block IIM, is being further developed due to the great value of this microsatellite.[110] Another small satellite development the U.S. Army is also involved in is the Army Resilient Global On-the-move SATCOM (ARGOS).[111] This LEO communication satellite system's main mission is to improve communications with the warfighter on the ground by using UHF via Army tactical radios, VHF, and beyond-the- horizon communications to guarantee reliable communications and support data

---

[106] Sandra Erwin, "Army Secretary: Still Unclear What Portions of the Army Would Move to the Space Force," *Space News*, last modified October 8, 2018, https://spacenews.com/army-secretary-still-unclear-what-portions-of-the-army-would-move-to-the-space-force/.

[107] "Future Warfare Center," U.S. Army Space and Missile Command, accessed Jun 1, 2019, https://www.smdc.army.mil/ORGANIZATION/FWC/.

[108] Johnson and Lindsey, 4.

[109] Sandra Erwin, "Army's Imaging Satellite Up and Running, but Its Future Is TBD," *Space News*, last modified February 21, 2018, https://spacenews.com/armys-imaging-satellite-up-and-running-but-its-future-is-tbd/.

[110] Erwin, "Army's Imaging Satellite."

[111] Johnson and Lindsey, 6.

exfiltration in various scenarios.[112] The ARGOS demonstrator satellites are projected to be launched around 2020.[113]

With the 1st Space Brigade being the primary space entity within ARSTRAT/SMDC,[114] its mission is to provide continuous space support operations to end users.[115] To accomplish this, many entities are required to support, including the Army Space Support Teams. Ultimately subordinate to ARSTRAT/SMDC,[116] they provide space capabilities, such as intelligence, surveillance, and reconnaissance, satellite communication, missile warning, and a host of other monitoring resources globally with impressive timing.[117]

The 53rd Signal Battalion (SATCON), who falls subordinate to the 1st Space Brigade in mission control,[118] also falls under the ARSTRAT/SMDC chain of command. It is located in Colorado, with the main mission of providing "satellite transmission, payload control, and electromagnetic interference detection of the Wideband Global System (WGS) and the Defense Satellite Communications Systems (DSCS)."[119] SATCON is the only unit that conducts payload and transmission control for the SHF wideband systems WGS and DSCS.[120] The Battalion is currently operating five Wideband

---

[112] Johnson and Lindsey, 6.

[113] Johnson and Lindsey, 7.

[114] Department of the Army, *Army Space Operations,* FM 3-14 (Washington, DC: Department of the Army, 2014) 6–1, https://fas.org/irp/doddir/army/fm3_14.pdf.

[115] Department of the Army, 6–1.

[116] Department of the Army, 1–7.

[117] Kevin T. Campbell, "The Army's Space Cadre," *High Frontier*, November 2007, vo. 4 no. 1,4,https://documents.theblackvault.com/documents/space/AFD-071119-017.pdf; USASMDC/ARSTRAT, "Army Space Support Teams," U.S. Army, last modified March 23, 2017, https://www.army.mil/standto/2017-03-23.

[118] Department of the Army, 6–1.

[119] Department of the Army, 6–8

[120] Jason B. Cutshaw, "53rd Signal Battalion Leads the Army's Space Operations," U.S. Army, last modified November 29, 2017, https://www.army.mil/article/197230/53rd_signal_battalion_leads_the_armys_space_operations.

Satellite Communications Operations Centers (WSOC), and a DSCS certification facility and also provides backup for the Air Force's Satellite control Network which will be discussed later through the WSOC.[121]

This control is vital to support unremitting communications on a tactical, operational, and strategic level. Col. Richard L. Zellman, former 1st Space Brigade Commander states that the "first line of defense should an adversary attempt to disrupt the SATCOM links provided by the WGS and DSCS satellite constellations."[122] On March 15, 2019, the 10th satellite joined the WGS constellation.[123] This high-capacity military communications satellite was launched from Cape Canaveral, Florida, to closer achieve the ever- constant goal of increased bandwidth capacity to support the warfighter in speedy decision making.[124]

E.   CONCLUSION

With a rich history in rocket development, thanks in part to captured German expertise, and currently taking particular interest in small satellite developments, the U.S. Army has proven reliability in the space game. Lt. Col. Glenn O. Mellor, former 53rd Signal Battalion commander stated that "more than 70 percent of weapons and equipment in the Army have a degree of reliance on a space-based service or capability."[125] Greater understanding of the energetic particle environment in space through further understanding of the Van Allen radiation belts has come about indirectly because of the U.S. Army and its role in space. Though the Army's focus shifted during the Vietnam War, there was always traction within it for space. The Army still views space and all space-trained personnel as key to the fight. U.S. Army Vice Chief of Staff Gen. James McConville stated

---

[121] Department of the Army, 6–8.

[122] Cutshaw, "53rd Signal Battalion."

[123] Sean Kimmons, "Satellite Launch Enhances Comms for Army Missions," U.S. Army, last modified March 18, 2019, https://www.army.mil/article/218701/satellite_launch_enhances_comms_for_army_missions.

[124] Kimmons, "Satellite Launch."

[125] Cutshaw.

in his answers to the Senate Armed services Committee that if a Space Force were to arise, only 500 out of the 2,500 space associated personnel would transfer over.[126]

The Army's restructuring of offices proved that space was important to the service and that it wanted to get things right. Eventually, the consolidation to ARSTRAT/SMDC was a good decision because now a direct line connects the U.S. Army to USSTRATCOM and now USSPACECOM. Because of the Army's involvement in space, and how important it is to every Army mission, the odds are very low that the Army will relinquish anything in space if the Space Force is approved.

---

[126] Erwin, "Majority of Army's Space Soldiers."

THIS PAGE INTENTIONALLY LEFT BLANK

III. A HISTORY OF THE NAVY'S ROLE IN SPACE AND CURRENT ACTIVITIES

A.   INTRODUCTION

This chapter assesses the history of the Navy's involvement in space. The purpose of this chapter is to give a clear picture of why the Navy became interested in space, and review what its current role is. This chapter first discusses the Navy's initial space involvement at the end of WWII and the evolution after the 1961 DoD decision to make the Air Force the central point of reference when it comes to space. This serves as a basis for assessing space power constraints within the Navy in relation to the current debate over the Space Force, and for providing recommendations while investigating policy implications.

B.   THE NAVY'S INITIAL SPACE INVOLVEMENT (1940s-1970s)

Victory in World War II came at the hands of the Grand Alliance among Great Britain, the Soviet Union, and the United States. Upon victory, the latter two emerged as great powers on the world stage for the first time, and began to compete with each other to captivate the minds and hearts of the rest of the world. There were not only two nations at odds with each other, but rather two ideologies that were fighting for supremacy: capitalism versus communism.[127] It was not only a conflict between Truman and Stalin, but the minds of Karl Marx and Friedrich Hegel were lined up against the genius of Adam Smith and Friedrich Hayek. Each power wanted to prove to the world that its beloved ideology was the way of the future for great achievement. This led to mini competitions between the two world powers, one of which was the space race. American author William E. Burrows articulates this perfectly by saying that "the Cold War would become the great engine, the supreme catalyst that sent rockets and their cargoes far above Earth and worlds away. If

---

[127] William E. Burrows, *This New Ocean: The Story of the First Space Age* (New York: Random House Publishing Group, 1998), 147.

Tsiolkovsky, Oberth, Goddard, and others were the fathers of rocketry, the competition between capitalism and communism was its midwife."[128]

Due to the success of Operation Paperclip[129] at the close of the war, America would seemingly come out as a winner by securing the services of Von Braun and others, along with missile parts that remained at their labs. The U.S. Army was able to work with General Electric to put together about 80 V-2 rockets, and offered some to the Navy. With the rockets and their engineers in hand, the U.S. wasted no time getting started on its road to space. By 1945, "each were using the captured... rockets for atmospheric sounding."[130]

1.   Launching of Vanguard and NRL Importance

The Navy spent the next ten years gathering information from sounding rockets that do suborbital trajectory launched into the earth's atmosphere. During this time, educational research was set up through the Naval Research Lab (NRL) and the Applied Physics Lab (APL) to explore the space realm. The NRL sent up 67 V-2 rockets to do such probing, then built the Viking rocket as a replacement; this became the precursor to the Vanguard rocket.[131] Simultaneously, the APL was working on launches with a liquid rocket it had developed called the Aerobee rocket.[132] These two rockets would later be used for satellite launches.[133] During this time, the Navy was also preparing for earth-orbiting satellites. In 1945, through the Navy's Bureau of Aeronautics, studies began on artificial earth satellites, and even a proposal for development was submitted the same year.[134] At the time, the

---

[128] Burrows, 147.

[129] Gary A. Federici, *From the Sea to the Stars: A Chronicle of the U.S. Navy's Space and Space Related Activities, 1944–2009* (Washington, DC: Department of the Navy, 2010), 8, http://www.history.navy.mil/books/ space/FromTheSeaToTheStars-2010ed.pdf.

[130] Federici, 8–9.

[131] Federici, 11.

[132] Constance M. Green and Milton Lomask, *Vanguard: A History* (Washington, DC: Office of Technology Utilization, 1970), 6, https://history.nasa.gov/SP-42-2.pdf.

[133] Green and Lomask, 16.

[134] Green and Lomask, 6.

Navy could not get on the same page with the Army's Air Corps (today's Air Force), making several attempts to work with them on building a satellite, but nothing materialized.[135]

Finally, the Army and Navy figured out the particulars to work together on a satellite concept, Project Orbiter.[136] The plans were cut short, however, because the Navy decided to work on its own project, Project Vanguard, due to the International Geophysical Year (IGY).[137] Influential scientists of the time wanted to have a year solely dedicated to a worldwide scientific endeavor; they settled on 1957–1958.[138] In 1955, the White House decided that the U.S. contribution for that year would be a satellite launch.[139] The Soviet Union made its own proposal in June of 1957 for the launch of a satellite.[140] With a two-year notice head start, the nation then was highly surprised when the Soviet Union launched *Sputnik* just four months later. A month after their first successful launch, the Soviets launched *Sputnik 2*.[141]

The Soviets' success lit a fire under the nation. In December of 1957, the Navy attempted with project Vanguard to launch the Viking rocket, which was scheduled to send off the country's first artificial satellite into Earth's orbit, but left the nation embarrassed as it blew up on the launch pad.[142] The Navy did eventually redeem itself after the Army

---

[135] Green and Lomask, 6.

[136] Green and Lomask, 14.

[137] Green and Lomask, *10.*

[138] Green and Lomask, 14.

[139] Green and Lomask, 14.

[140] Green and Lomask, 15.

[141] Federici, 15.

[142] Federici, 15.

did damage control by successfully launching a three and a half pound satellite into orbit, *Vanguard 1,* on March 17, 1958.[143]

Even with this success, it was recognized that U.S. national security was still being threatened and, thus, the NRL created the Navy Space Surveillance System (NavSpaSur) to identify and track foreign satellites.[144] The NRL has been important in the development, testing, and supervision of many projects. The Galactic Radiation and Background (*GRAB*) satellite was also a project managed by the NRL/NRO (National Reconnaissance Office) and launched in June of 1960. This Navy electronic intelligence (ELINT) satellite system earned the label of the world's first intelligence, surveillance, and reconnaissance (ISR) Satellite. The *Grab* program gained Soviet air defense radar intelligence that the other services were not able to collect.[145] It was also the precursor to programs such as the *Poppy* satellite. During this time, the Navy saw satellite technology as essential, and by 1960 already had developed the Navy Navigation Satellite System (NNSS), also known as the *Transit.*[146] It was the very first Navigational system to be used operationally and a trailblazer for systems we now rely heavily on today. However, in 1961, the DoD made a decision that changed the direction of who controlled the satellites for a little less than a decade.

2. The 1961 Department of Defense Directive 5160. 34

The 1961 Department of Defense Directive 5160.34 (DoDD 5160.32) gave governing responsibility to the U.S. Air Force for "1. Research, development and operation, including payload design, launch, guidance, control and recovery of all DoD reconnaissance satellite systems. 2. Research and development of all instrumentation and

---

[143] Division on Engineering and Physical Sciences, *Navy's Needs in Space for Providing Future Capabilities* (Washington, DC: National Academy of Sciences, 2005), 152, https://www.nap.edu/read/11299/.

[144] Division on Engineering and Physical Sciences, 152.

[145] Robert A. McDonald and Sharon K. Moreno, Raising the Periscope...Grab and Poppy: America's Early ELINT Satellites (Chantilly, VA: Center for the Study of National Reconnaissance, 2005), 14, https://www.nro.gov/Portals/65/documents/history/csnr/programs/docs/prog-hist-03.pdf.

[146] Boehm, 22.

equipment for processing reconnaissance data from satellite sources"[147] With this in effect, it gave no allowance for the Navy to implement or launch any new satellites, but it did continue the research and advancement of satellites, and also other testing areas that would advance the space realm. For example, in 1962, the Navy Astronautics Group (NAVASTROGRU), later known as the Naval Satellite Operations Center (NAVSOC), was created to operate the *Transit* satellite.[148] It is also responsible for monitoring the spectrum frequencies, which are still vital to global positioning satellites (GPS), and operating other communication satellites such as UHF SATCOM that are still in use today.[149]

The Navy also played a role in anti-satellite weapons (ASAT) due to a concern for space defense and the ultimate desire of U.S. space supremacy. In the early 1960s, the Navy formed the high altitude program (HI-HO/HI-HOE) test in which it launched missiles from the air that left the atmosphere before hitting the target.[150] This program, at the time secret in classification, was new and innovative, and though its last launch was in 1962, it laid the groundwork for future ASAT testing.[151]

In September of 1970, the DoD decided there was strength in allowing each of the services to be able to continue in the development of satellites for specific warfare needs and capabilities, so DoDD 5160.32 was revised.[152] Though the Air Force was still

---

[147] Department of Defense, *Reconnaissance, Mapping and Geodetic Programs Reference*, DoD Directive 5160. 32 (Washington, DC: Department of Defense, 1961), https://www.nro.gov/Portals/65/documents/foia/declass/WS117L_Records/215.PDF.

[148] NHCC, "Navy Astronautics Group Established, 22 May 1962," Naval History (blog), U.S. Naval Institute, May 22, 2010, https://www.navalhistory.org/2010/05/22/navy-astronautics-group-established-22-may-1962.

[149] Clint W. Miller, "Optimizing the Navy's Investment in Space Professionals" (space master's thesis, Naval Postgraduate School, 2011), https://calhoun.nps.edu/bitstream/handle/10945/5513/11Sep_Miller_C.pdf?sequence=1&isAllowed=y.

[150] "HiHo / Hi-Hoe / NOTS-EV-2 Caleb," Global Security, https://www.globalsecurity.org/space/systems/hiho.htm.

[151]"HiHo / Hi-Hoe / NOTS-EV-2 Caleb."

[152] Department of Defense, *Development of Space Systems*, DoD Directive 5160. 32 (Washington, DC: Department of Defense,1970), https://apps.dtic.mil/dtic/tr/fulltext/u2/a272407.pdf

established as the leader for space development,[153] all of the services were now allowed to contribute in a greater way than just to research.

C. THE NAVY'S SPACE ADVANCEMENT (1980s–PRESENT)

1. Space Reorganization within the Navy

The Navy's technology advanced rapidly as more ways were figured out how to utilize instruments of space to further objectives on the earth's surface. The 1980s were a time of a rise in space organization in the U.S., as Ronald Reagan entered the presidential office and had a desire to see growth in the military, defense spending, and space presence.[154] One of his proposals was the Strategic Defense Initiative (SDI), an initiative that would protect the U.S. via space from the Soviet Union's nuclear missile attacks. Though SDI did not emerge, various research and programs did.[155] Many organizational changes also happened in the 1980s within the Navy, one being the stand-up of a space curriculum at the Naval Postgraduate School (NPS) in 1982. There was a major emphasis on educating young officers so they would return to their operations having knowledge of the space domain. From an operational perspective, in October of 1983, the Naval Space Command (NAVSPACECOM) was stood up in hopes of consolidating any developments or organizations that dealt with space ventures within the Navy.[156] Thus, it gained operational control (OPCON) of systems like NavSpaSur. In 1985, NAVSPACECOM was combined with the United States Space Command (USSPACECOM) and was positioned under the command as the primary naval element. This move aimed at the consolidation of space activities amongst the Navy as well as the other services for a more united and clear space front.

---

[153]Boehm, 23.

[154] "Ronald Reagan`s Military Buildup," United States History, https://www.u-s-history.com/pages/h1957.html.

[155] "Ronald Reagan`s Military Buildup."

[156] Boehm, 22.

2. Space Refocus within the Navy Post-Cold War

After the Cold War, there was a change in attitude concerning the amount of resources that should be dedicated to space, especially within the minds of Navy higher-ups.[157] The attitude from the top was mostly concerned with tactical strategy and maneuvers within Earth's orbit. There seemed to be a gap in reasoning, though, because the Navy utilized space technology for the day-to-day operations of ship communications, navigation, weather and so on, which did help with operations within Earth's orbit. Also, their attitudes mostly reflected the challenges of the time as well. The new fight was not with the Soviets in space, but in the Middle East against Iraq.

In the early 1990s, a press toward space resource allocation to other areas began, but there was still a faction within the Navy that desired a space presence. In 1997, the first "Smith" panel,[158] named after Admiral William D. Smith, addressed the issues of the languishing Navy interest in space, as well as the reorganization of the NRO. The panel forwarded a report to the Secretary of the Navy and the Director of the NRO with a few insights, highlighting the value of Navy personnel to the NRO, the high operational return from the personnel involved, a recommendation to commit full resources necessary for a Navy-NRO partnership, and a recommendation for the previously named Space and Naval Warfare Systems Command (SPAWAR) and NRO consolidation under one flag.[159] This never happened, as America's focus was redirected to more important issues of the time.

With an increasing attention to the fight against terrorism after 9/11, USSPACECOM was dissolved in 2002, and NAVSPACECOM went along with it. The reason behind the disbandment of USSPACECOM was because of the stand-up of the United States Northern Command (USNORTHCOM) after the 9/11 attacks. The country's focus now had very little to do with space and everything to do with protecting itself from another terrorist attack. Because of this, the previous responsibilities USSPACECOM had

---

[157] Federici, 109.

[158] Federici, 143.

[159] Federici, 141.

33

were transferred over to U.S. Strategic Command (USSTRATCOM). Within USSTRATCOM, the Combined Space Operation Center (CSpOC) reports to the Joint Functional Component Command for Space now known as the Joint Force Space Component Commander and executes C2 of space assets 24/7.[160] Also, a significant portion of responsibility went to the Air Force and any specific naval space operations fell under the responsibility of SPAWAR.[161] Though USSPACECOM was disbanded, the U.S. Navy stood up the Naval Network Warfare Command in 2002, which is the Navy's current operational control for space.[162] The "Navy Space Policy Implementation" released by the Deputy of Chief Naval Operations (DCNO) in 2005,[163] gives SPAWAR responsibility for the support of the Navy space architecture, in coordination with the Director of Naval Intelligence.[164] It also lays out the official Navy Space Cadre (NSC) designator.[165]

With the Navy's main goal to "use space to gain and maintain Information Superiority in naval operations,"[166] the official formalization of the NSC in April of 2004 was fitting.[167] Since June of 2009, the NSC has fallen under the Information Dominance Corps (IDC), now formally known as the Information Warfare Community (IWC) within

---

[160] "U.S. Strategic Command Fact Sheet Combined Space Operations Center /614th Air Operations Center," USSTRATCOM, July 2018, https://www.stratcom.mil/Portals/8/Documents/CSpOC_Factsheet_2018.pdf.

[161] Boehm, 22.

[162] "Navy Establishes Naval Network Warfare Command," U.S. Navy, last modified March 28, 2002, https://www.navy.mil/submit/display.asp?story_id=1156.

[163] Department of the Navy, Navy Space Implementation, DoD Instruction 5400.43 (Washington, DC: Office of the Chief of Naval Operations, 2005), 1.

[164] Department of the Navy, 5.

[165] Department of the Navy, 1.

[166] Boehm, 22.

[167] Matthew E. Faulkenberry, "Critical Review of the Navy Space Cadre" (space master's thesis, Naval Postgraduate School, 2014), 20, https://www.hsdl.org/?view&did=762129.

the Navy.[168] VADM Brian Brown, the current head of the NSC,[169] stated that "space expertise in our ranks is critical to success in the maritime domain."[170]

The current space programs the Navy has OPCON over are the Ultra-High Frequency Follow-On (UFO), and its successor, MUOS. Currently, a new MUOS wideband code division multiple access (WCDMA) payload is planned to be fully operational late 2019.[171] This will provide MUOS users "simultaneous voice, video, and data capability by leveraging 3G communications technology."[172] The UFO, with the predecessor being Fleet Satellite Communications (FLTSATCOM), is a constellation of eleven geosynchronous communications satellites. Of the eleven, five are still operational.[173] This constellation provides priority strategic communications for warfighters. The Navy Communications Satellite Program Office (PMW 146) has managed this program to provide operational support to the DoD and other entities both domestic and abroad.[174]

The Navy has recently expressed the view that information spreads and is relatable across the warfare areas of sea, land, air, and space. Because of this, SPAWAR was changed to Naval Information Warfare Systems Command (NAVWAR), effective June 6,

---

[168] Faulkenberry, 20.

[169] Brian B. Brown, "2018 Retrospective and Look Ahead For 2019" (official memorandum, Washington, DC: United States Navy, 2019), 1, https://www.mnp.navy.mil/documents/34084/0/Space+EOY+Summary+for+2018.pdf/290f72db-31e7-0646-914d-51ad2972e137.

[170] Brown, 1.

[171] SPAWAR, *2018 The SPAWAR List* (San Diego, CA: SPAWAR, 2018), 38, https://www.public.navy.mil/navwar/Documents/List.pdf.

[172] SPAWAR, 38.

[173] Benjamin D. Forest, "An Analysis of Military Use of Commercial Satellite Communications," (master's thesis, Naval Postgraduate School, 2008), 28, https://core.ac.uk/download/pdf/36697715.pdf.

[174] "PEO Space Systems Fact Sheet," Department of the Navy, accessed February 26, 2019, https://www.public.navy.mil/navwar/PEOSpaceSystems/Documents/146_FactSheet_MUOS5_%202016_FINAL_7Oct16.pdf.

2019.[175] The name change was due to an effort to "normalize information warfare into the way we do operations and warfighting in the Navy,"[176] said Chief of Naval Operations John Richardson. This new name is said to more "accurately describe the mission."[177] Though NAVWAR still supports space endeavors, the name change is due to a desire to view space from a broader standpoint. NAVWAR supports "naval warfare from seabed to space."[178]

Without the different advantages space brings, the Navy would be nowhere close to the fighting force it is now. There was much incorporation of space into exercises in the year 2018 such as Global Lightening (USSTRATCOM's yearly exercise), Fleet Design Wargame (held at the Naval War College in Newport, RI), and more.[179] The NSC is increasingly receiving space-coded billets that help surge space knowledge to the operational level.[180] The number of personnel becoming qualified in the NSC is currently rising. In 2018, 399 students completed the Naval Space Operations Course (NSOC) with the largest class of 73 of that number.[181] Other training courses such as Space 200/300 and the Space Warfighter Preparation Course (SWPC) have also had an increase in the personnel taking them in.[182]

Finally, the NRL continues to be involved in different space oriented research. For example, the NRL is currently working along with other agencies to attack the increasing

---

[175]Barry Rosenberg, "Why Is SPAWAR Now NAVWAR? Networks & Cyber Warfare," Breaking Defense, June 5, 2019, https://breakingdefense.com/2019/06/why-is-spawar-now-navwar-networks-cyber-warfare/.

[176] John Richardson, "SPAWAR to NAVWAR," June 2, 2019, produced by Naval Information Warfare Systems Command, 2:00:00, https://www.youtube.com/watch?v=y1EJ-3har-I.

[177] Richardson, "SPAWAR to NAVWAR."

[178] Richardson.

[179] Brown, 3.

[180] Brown, 2.

[181] Brown, 2.

[182] Brown, 2.

problem of orbital debris.[183] The NRL is also working with the Air Force to complete the new space fence, the daughter of the NavSpaSur, a program that the Navy shifted to the Air Force in 2004.[184]

D.   CONCLUSION

Though the U.S. Navy has evolved since its initial participation in space, it has remained constant in supporting it. Even through structural changes, the Navy has always somehow kept space a priority. The Navy has been able to recognize throughout the years that the space domain affects all other warfighting domains and, thus, keeping it relevant in regard to the fight is important. It is also important to note that, though the Navy has viewed space as important, it has always kept it as a subset or support mission and never made it a primary mission. Even with the space cadre stood up, there is still an elusiveness within the Navy when it comes to space being a mission area of its own. Though it is a part of the Information Warfare community, it is not publicly recognized as such because it is not its own community. VADM Brian Brown stated in the "2018 Retrospective and Look Ahead for 2019" that "the Navy supports actions that integrate space across the air, sea, and land warfighting domains and move space governance closer to the warfighter."[185]

Though the Navy is expressing more interest in space, it is no more than a partner with certain mission areas for space, thus the importance of the name change from SPAWAR to NAVWAR. The Navy is also integrated within centers such as the CSpOC, formerly known as the Joint space operations center (JSpOC).[186] There are Navy space cadre billets on the CSpOC watch floor in which the personnel give situational awareness

---

[183] Emanuel Cavallaro, "Eyes on the Skies: NRL Researchers Tackle the Ever-Growing Problem of Orbital Debris," U.S. Naval Research Laboratory, https://www.nrl.navy.mil/space/eyes-skies-nrl-researchers-tackle-ever-growing-problem-orbital-debris.

[184] Cavallaro, "Eyes on the Skies."

[185] Brown, 1.

[186] Joint Force Space Component Command Public Affairs, "Combined Space Operations Center established at Vandenberg AFB," *Air Force Space Command*, July 19, 2018, https://www.afspc.af.mil/News/Article-Display/Article/1579285/combined-space-operations-center-established-at-vandenberg-afb/.

to satellite users and to leadership, as well as insight to the warfighters within the Navy as to the space capability that best fits various missions.[187] It was recently announced that the CSpOC will be a big support unit to the newly re-established U.S. Space Command (USSPACECOM).[188] The Navy is integral because of the involvement with satellite communications, especially with how fast the demand is growing for it. Right now, the CSpOC is mostly comprised of airmen and has about 50 dedicated joint positions out of 450 but it is expected to increase.[189] The other dedicated center that will be a heavy lifter in the support of USSPACECOM will be the National Space Defense Center (NSDC).[190] Though no direct work with the Navy has been established, these different entities are important to show the interconnectedness USSPACECOM is trying to achieve.

Thus, though the Navy has been in charge of some of the oldest space programs, it is far from being the lead for anything space related within the armed forces, due to its high focus on other warfare domains and reliance on the U.S. Air Force for space support.

---

[187] Adam DeJesus, "Navy Space Cadre & Warfighters Putting Pieces Together," *CHIPS, The Department of the Navy Information Technology Magazine*, June 10, 2014, https://www.doncio.navy.mil/chips/ArticleDetails.aspx?ID=5186.

[188] Sandra Erwin, "U.S. Space Command's Major Components Will Be Based in California and Colorado," *Space News*, June 30, 2019, https://spacenews.com/u-s-space-commands-major-components-will-be-based-in-california-and-colorado/.

[189] Erwin.

[190] Erwin.

IV. U.S. AIR FORCE ROLE IN SPACE ACTIVITY

A.   INTRODUCTION

This chapter discusses the history of the USAF role in space. It begins with an exchange amongst the leaders within the U.S. Army Air Forces that believed space flight to be possible and how this paradigm shifted after the USAF separated from the U.S. Army. The chapter then works through how the USAF eventually claimed the title of being the executive agent for space within the DoD. It also breaks down the different reorganizational challenges the USAF has had, including the cyber domain expansion and some of the USAF current missions and space resources. It then concludes with USAF integration through its leadership of the reestablished USSPACECOM and different USAF leadership opinions on the effectiveness of establishing a Space Force/Corps. This chapter will help shape the framework for the Space Force discussion in chapter 5, where I will be providing policy analysis and recommendations.

B.   EMERGENCE FROM U.S. ARMY AIR FORCES TO U.S. AIR FORCE: 1945–1961

U.S. Air Force (USAF) involvement in space dates back to before it became a separate service apart from the U.S. Army. The U.S. Army Air Forces (AAF) toward the end of WWII had many leaders that viewed spaceflight as possible. The AAF Commanding General Henry H. Arnold wrote in 1945 that "a space ship is all but practical today and could be built in the foreseeable future."[191] Shortly after, the Air Force Scientific Advisory Group noted that satellites and long-range rockets were both a "definite possibility."[192] All of the services were interested in Space, but AAF leadership believed that space operations should be solely handled by the Air Force.[193] In fact, in 1946, the Director of Research

---

[191] Curtis Peebles, *High Frontier: The United States Air Force and The Military Space Program* (Washington, DC: Air Force History and Museums Program, 1997), Intro, https://media.defense.gov/2010/Dec/02/2001329901/-1/-1/0/AFD-101202-013.pdf

[192] Peebles, Intro.

[193] Peebles, Intro.

and Development for the AAF, Major General Curtis E. LeMay, rejected a joint service space program proposal presented by the Aeronautical Board even after the AAF board members expressed interest.[194] LeMay then ordered an independent study to help the AAF establish dominance in the field.[195] For success with this study, they requested Project RAND for technical guidance.[196]

Project RAND was a special contractor under the Douglas Aircraft Company and officially began operating in December of 1945 with a mission to join military developments with advanced research.[197] On May 2, 1946, Project RAND released its first study, "Preliminary design of an Experimental World-Circling Spaceship."[198] This study discussed a prospective strategy for design and the use of man-made satellites, as well as conceivable roles for the technology such as communications and weather observation.[199] The study also went on to predict that the U.S could launch a "500-pound satellite into a 300-mile orbit for $150 million"[200] in five years. Project RAND went on after to conduct studies on whether these proposed satellite capabilities would be feasible headed by the Air Material Command.[201]

With space innovations underway, understanding of the need for airpower resources to be solely dedicated to a separate service outside of the U.S. Army was proceeding as well. The debate about the need for strengthening air power and an independent Air Force had been longstanding. Thus, on July 26, 1947, President Harry S.

---

[194]David N. Spires, *Beyond Horizons: A Half Century of Air Force Space Leadership,* (Washington, DC.: Air Force Space Commands, 1998), 14, https://apps.dtic.mil/dtic/tr/fulltext/u2/a355572.pdf.

[195]Spires, 14.

[196] Peebles, 1.

[197]"A Brief History of Rand," RAND Corporation, accessed August 9, 2019, https://www.rand.org/about/history/a-brief-history-of-rand.html.

[198]Peebles, 1.

[199]Peebles, 1.

[200]Spires, 15.

[201]Peebles, 1.

Truman signed the National Security Act, which reorganized many organizations and, most importantly, recognized the Army Air Corps as an independent service, the USAF.

1.   U.S. Air Force Struggle for Space

Once the USAF gained its complete independence as a separate force, the leadership pushed harder to be assigned administrative and operational control of "any future U.S. satellite and missile development."[202] The USAF was soon granted approval by the DoD to develop strategic missiles and satellites in 1948. The USAF continued studies such as Project WIZARD and Project THUMPER. Both ballistic defense studies, Project THUMPER was cancelled in March of 1948 and Project Wizard went on the merge with the Army's NIKE-ZEUS missile system. Also, in 1949, USAF planes collected radioactive fallout proving that the Soviet Union was conducting nuclear tests. This preceded the creation of the U-2 plane, primarily used for reconnaissance, but approved by President Dwight D. Eisenhower in 1954 to sample the upper atmosphere for indication of nuclear weapon testing.[203] This discovery helped Air Force leadership to realize that space missions would be very important. This was all expedient because of technical innovations making it more achievable to develop small nuclear warheads, which will be discussed later on.

Throughout it all, the USAF fight continued against the other services in regard to who would be assigned to different roles and operations of space development, and the USAF still desired to have exclusive rights to space. Four years earlier, in 1950, the Air Force was awarded responsibility for the development of long-range strategic missiles and short-range tactical missiles.[204] Also in 1950, the Air Research and Development Command (ARDC) was established,[205] and by 1953 it had taken over all of Project

---

[202] Benjamin S. Lambeth, *Mastering the Ultimate Higher Ground: Next Steps in the Military Uses of Space* (Santa Monica, Ca: RAND, 2003), 12, https://www.rand.org/content/dam/rand/pubs/monograph_reports/2005/MR1649.pdf.

[203] Peebles, 3–4.

[204] Lambeth, 13.

[205] Spires, 14.

RAND's satellite research studies.[206] ARDC found that consolidating all aspects of RAND satellite work into one project would be beneficial and easier in managing all of the work done. Project 409–40, the "Satellite Component Study" covered system development, but on December 3, 1953, ARDC headquarters altered the planning for system development into an actual proposed satellite reconnaissance system called the Weapon System-117L, which was later known as the Advanced Reconnaissance System.[207]

2. The Merging of Satellite and Missile Development

During the 1950s-early 1960s, the USAF also became involved with the advancement of intercontinental ballistic missiles (ICBMs).[208] Historian Walter Boyne argues that the Air Force's initial approach to space was "both curious and coincidental" developed by the need to make the ICBM. He argues this because as the time, the USAF did not have the resources or foresight for what the exploration or exploitation of space would entail.[209] The competition for resources between missiles and satellites was a fear for the USAF, and the Assistant Secretary for Research and Development during the time, Trevor Gardner made known that he was worried the Atlas ICBM schedule could conflict with satellite progress and requirements.[210] He shared his concern with the ICBM Scientific Advisory Group, headed by mathematician and physicist John von Neumann. In 1955, von Neumann recommended that satellite work be restricted to only the spacecraft to ensure the ICBM program would have resources and the room to accelerate in response to the Soviet Union and its nuclear warhead developments.[211] However, as work progressed on the ICBM, von Neumann and the Scientific Advisory group realized that the

---

[206]Joshua Boehm, "A History of United States National Security Space Management and Organization," Commission paper, Commission for Assessment of United States National Security Space Management and Organization, 2001, 25, https://fas.org/spp/eprint/article03.html#ft78.

[207] Boehm, 25.

[208] Boehm, 25.

[209] Lambeth, 10.

[210] Spires, 37.

[211]Boehm, 26.

two were very much interconnected, since the warhead was at least a temporary satellite.[212] In 1954, the USAF created the Western Development Division (WDD) under the ARDC, under the leadership of Brig. Gen. Bernard A. Schriever. It was tasked with the development of the Atlas as the primary ICBM, and as a backup, the Titan. Brig. Gen. Schriever was convinced that all military missile and satellite programs needed to be under integrated administration to fix the problem of constant competition for resources.[213] If it fell under the same roof, it would be easier to see how the problems were complimenting each other, not competing. Thus, in February of 1956, after consulting with Simon Ramo of Ramo-Wooldridge, WDD's technical consulting firm, all management for missile and satellite development was centralized.[214] As a result of this, Weapon System-117L was relocated to the WDD. Both were expected to thrive being at a command devoted to the missile and satellite space environment.[215]

The future of the Weapon System-117L remained in limbo for the next couple of years. In 1956, Lockheed Martin won the contract with the 117L reconnaissance satellite.[216] Though the development plan was approved, the DoD showed its lack of support by cutting the budget drastically from the originally requested $115 million to only $4 million for this endeavor.[217] By late 1956, the program was considered a "low priority, long-term effort,"[218] and even ICBM development faced restrictions until concern was reignited after the 1957 Soviet *Sputnik* launch.[219] In January of 1958, the National Security Council prioritized the development of a reconnaissance satellite as the highest priority for

---

[212]Spires, 37.

[213]Spires, 37.

[214]Spires, 37.

[215]Spires, 38.

[216]Peebles, 7.

[217]Boehm, 26.

[218] Peebles, 9.

[219] Boehm, 26.

the nation alongside U.S. ballistic missile systems.[220] With President Eisenhower now viewing space development as imperative, he commissioned the Advanced Research Projects Agency (ARPA) to manage defense research and advancement. This development also included military satellite communications.[221] When the National Aeronautics and Space Administration (NASA) was created, all civilian space activities were consolidated within it, and all military space activities were overseen by ARPA. Though ARPA had oversight, the USAF were still involved in missions such as Project Able, the unsuccessful attempt to put a satellite around the moon.[222] By the end of 1959, the 117L system had developed into three programs, the Discoverer Program (secret code name: CORONA), the Satellite and Missile Observation System (SAMOS), and the Missile Defense Alarm System (MIDAS).[223] SAMOS and MIDAS were somewhat short lived programs due to technical problems. However, research on these platforms helped construct future systems.

3.   The CORONA Program and NRO Establishment

The CORONA Program was authorized by President Eisenhower and jointly managed, like the U-2 plane, by the Central Intelligence Agency (CIA) and the USAF. CORONA's cover to the public was a scientific research program named Discoverer.[224] After multiple failures and high leadership discussion of the potential cancellation of CORONA, on August 10, 1960, Discoverer XIII was launched into orbit and two days later the recovery capsule was sent back with the American Flag. When Corona's program officer Major Ralph J. Ford sent the encrypted message of "Capsule recovered

---

220 Boehm, 26.

221 Peebles, 10.

222 "Project Able," Living Moon, accessed August 12, 2019, https://www.thelivingmoon.com/45jack_files/03files/Project_Able.html.

223 "Satellite Systems," accessed August 12, 2019, https://www.losangeles.af.mil/Portals/16/documents/AFD-060912-025.pdf?ver=2016-05-02-112847-777.

224 "Discoverer/ Corona: First U.S. Reconnaissance Satellite," Smithsonian National Air and space Museum, accessed August 14, 2019, https://airandspace.si.edu/exhibitions/space-race/online/sec400/sec420.htm.

undamaged,"[225] there was definitely celebration after more than eighteen months of failures. Only one week later, on August 18, 1960, Discoverer XIV carried a camera into orbit and the recovery capsule was sent back with the first U.S. reconnaissance satellite photos of Soviet terrain taken from space.[226] President Eisenhower decided that it was best not to aggravate the Soviets by releasing the pictures to the public and, thus, they were kept classified. The CORONA program was active all the way up until 1972, with 145 flights under its belt.[227]

When President John F. Kennedy took the Presidential office in January of 1961, he validated the existing space missions and developments. The year of 1961 also brought the establishment of the National Reconnaissance Office (NRO), which ended the USAF's operational control over satellite reconnaissance programs.[228] From now on, the USAF would simply deliver the satellite to orbit and return the film capsule.

C. USAF EXECUTIVE CONTROL AND REORGANIZATION INTO A UNIFIED COMMAND FOR SPACE: 1961–1985

With NASA taking shape and a decent part of the Army and Navy space programs being dissolved into it, for the first time the military deemed the USAF as "the front-runner for the military space mission."[229] With the support of Secretary of Defense Robert S. McNamara, the year 1961 saw some reorganizations within the USAF to further support research and development for space systems. Initially the Air Research and Development Command, the USAF re-designated this command in 1961 to be the Air Force Systems Command (AFSC) and set up in four divisions under General Bernard Schriever to manage space and missile systems more proficiently.[230] With the DoDD 5160.32 signed off the

---

[225]Theresa Foley, "Corona Comes in from the Cold," *Air Force Magazine*, last modified September 1995, http://www.airforcemag.com/MagazineArchive/Pages/1995/September%201995/0995corona.aspx.

[226] Peebles, 14.

[227] Peebles, 14.

[228] Boehm, 26.

[229] Spires, 66.

[230] Boehm, 26.

Air Force saw itself as the head of military space development.[231] USAF leadership realized that having a good working relationship with NASA would be expedient for them, especially since the USAF was assigned responsibility for launching all space boosters as well as the integration of the payloads, taking the control of space system development back from ARPA.[232] With NASA heavily depending on the USAF for these space boosters, and the USAF being responsible for more that 90 percent of all military efforts dealing with space,[233] the DoD established the USAF as "the executive agent" for NASA support through DoDD 5030.18 in 1962.[234]

During the 1960s, the USAF was involved in many space advancements. The first military communication satellite system (Initial Defense Communications Satellite Program), used for operations was put under the USAF in 1962.[235] The program known as Dyno-Soar was re-oriented in 1962 into the X-20 Dyna-Soar. This program, once created to enhance military missions through a spaceplane, would now be solely used for research. The X-20 Dyna-Soar was a leader for programs such as NASA's Gemini. Though the X-20 was cancelled in 1963, the tests it conducted of reentry technology would be carried further by the ASSET program,[236] a sequence of small re-entry vehicles. Secretary McNamara wanted the USAF to participate in manned flights.[237] Blue Gemini[238] was a project started in 1962 to enable the USAF to develop manned spaceflight. At a meeting with NASA in late 1962, McNamara proposed that the two Geminis (NASA and the USAF)

---

[231] Department of Defense, *Reconnaissance, Mapping and Geodetic Programs Reference*, DoD Directive 5160. 32 (Washington, DC: Department of Defense, 1961), https://www.nro.gov/Portals/65/documents/foia/declass/WS117L_Records/215.PDF.

[232] R. Cargill Hall and Jacob Neufeld, *The U.S. Air Force in Space: 1945 to the Twenty-first Century* (Washington, DC: USAF History and Museums Program, 1995), 41, https://media.defense.gov/2010/Oct/01/2001329745/-1/-1/0/AFD-101001-060.pdf.

[233] Boehm, 27.

[234] Boehm, 27.

[235] Spires, 23.

[236] Peebles, 21–22.

[237] Peebles, 20.

[238] Hall and Neufeld, 72.

be merged together, but NASA rejected this idea.[239] The Blue Gemini project was cancelled in the late 1960s because of funding issues. However, McNamara did not stop there. He saw it fitting that with manned space flight efforts; a space station would be helpful for research. In 1963, Harold Brown who was the soon to be Director of Defense Research and Engineering, approved the USAF to do a study of a space station.[240] With all the other space developments taking place and with AFSC leadership believing that the USAF needed another entity to become more effective in different space ventures, such as payload recovery and launching authority, the National Range Division was established in 1964.[241] This division helped to coordinate launch activity responsibility with both the DoD and NASA. A couple of years later, in 1967, the AFSC developed the Space and Missile Systems Organization (SAMSO), which consolidated all USAF missile and space functions. A successor to this organization was the Aerospace Defense Command (ADCOM), established in 1968 for the actual monitoring of missile launches.[242] For the space station undertakings, NASA stated that it was not able to support the recommended USAF space station and proposed a manned laboratory. This dissipated worries of resupplying the station, and crew docking and transfer by proposing that a 1,500-cubic-foot laboratory be attached to a Gemini and launched by a Titan IIIC with the laboratory being functional for 30 days then being released to deorbit.[243] This design, influenced by NASA, was called the Manned Orbiting Laboratory (MOL). Though the USAF was set to direct this program, it had no involvement with the decision making progress regarding it. Even though plans for the MOL development were submitted, the program goal was redirected to manned space reconnaissance due to medical experts confirming that "humans could not survive for 30 days in a weightless orbital environment."[244] To support

---

239 Peebles, 20.

240 Peebles, 20–21.

241 Boehm, 27.

242 Boehm, 27.

243 Peebles, 21.

244 Peebles, 23.

the reconnaissance mission, the launch site was moved from Florida to California so that MOL could be in a polar orbit.[245] The first 5 MOL launches were originally scheduled to launch in 1969, however, the program dealt with major technical and budget issues,[246] and in that year, the initial cost of the MOL program had been doubled to $3 billion and still there had been no launch. MOL was eventually cancelled in an effort to reduce military spending.[247]

1.   USAF Operational Organization Efforts for Space

With the need for space becoming more crucial, and the military and DoD able to see how space systems sustained vital operations through satellite communications and photoreconnaissance during the Vietnam War,[248] the USAF grew even more serious about the need for an organization dedicated solely to space. In 1974, the ADCOM Commander sent a letter to the USAF Chief of Staff detailing an opportunity for an improved space organization within ADCOM. Guidance from different Air Force major commands led to a 1977 space policy study to get a detailed look at all of the space missions associated with the USAF, which became known as "The Navaho Charts" because of "the many colors associated with the different functions and systems."[249] This highlighted the complexities of space operations and the different organizations associated with them. This situation prompted General David Jones, the USAF Chief of Staff at the time, to conduct a special meeting in which he proposed to remove ADCOM and North American Aerospace Defense Command (NORAD) from all organizational structures. However, the group came to the conclusion that because of Canadian influence within NORAD, it needed to be kept. But in 1979, ADCOM was discontinued and its different assets were divided among the commands and organizations. The plan on how to divide was outlined in "the Green Book,"

---

[245] Peebles, 23.

[246] Hall and Neufeld, 71.

[247] Peebles, 26.

[248] Spires, 169.

[249] Hall and Neufeld, 136.

the product of a study committee tasked to recommend where the different assets would go.[250] After the ADCOM's disestablishment, the USAF's quest for space consolidation continued in the 1978 "Space Missions Organizational Planning Study" (SMOPS). Written by a space advocacy group at the Air Staff, this study proposed five structural options within the USAF that could potentially help with space organization. The recommendations included a possible USAF command for space and the appointment of the USAF as the executive agent for space within the DoD.[251]

2.   The Creation of the Air Force Space Command

The early 1980s brought a rude awakening to the USAF as a study done by the Air Force Scientific Advisory Board determined that the USAF was ill prepared for space due to an unclear organizational structure. This study resulted in AFSC appointing Major General Jack Kulpa as the Space Division's Deputy Commander for Space Operations who dual hatted the same role at SAMSO.[252] Major General Kulpa also directed and presented a final report briefed to the Secretary of the Air Force on November 4, 1980. This report educated leadership on essential capabilities needed for the creation of a Consolidated Space Operations Center (CSOC) within the AFSC.[253] In 1981, the USAF established the CSOC in Colorado Springs, Colorado, but would move it to Schriever Air Force Base, about 25 minutes east.[254] Sequentially, LT. General O'Malley became the XO on the Air Staff and called for a devoted space command. He stated that this office would standardize space operations within the USAF by "providing a renewed emphasis that the Air Force

---

[250] Hall and Neufeld, 137.

[251] Hall and Neufeld, 72.

[252] Boehm, 27.

[253] J. Catherene Wilman, *Space Division: A Chronology, 1980–1984* (Office of History Headquarters, 1991), 45, https://books.google.com/books?id=mXEgedvuspMC&printsec=frontcover&source=gbs_ge_summary_r&cad=0#v=onepage&q&f=false.

[254] Tom Roeder, "Space Force: A Timeline," *The Gazette*, last modified June 25, 2018, https://www.coloradopolitics.com/news/space-force-a-timeline/article_307d061b-687c-5332-8286-9b5556c5be61.html.

plans to stay in the lead in military space operations."[255] Major General Jack Chain worked with General O'Malley as the Pentagon liaison in creating a proposal for why space activities in the USAF should be under a separate command.[256] He stated that he would call it "the Fester Briefing," to make sure it festered on the desk of General O'Malley until he acted upon it.[257] In 1981, a Directorate for Space Operations was formed to focus on space affairs within O'Malley's office.[258] Simultaneously in 1981, President Ronald Reagan took office and released the Strategic Modernization Plan. Though this plan was intended to discuss U.S. influence on space, it ended up calling the USAF to develop a "reorganizational roadmap for future space activities."[259] Major leadership was all for this plan moving forward, which prompted a call for the USAF to create a major command for space operations to achieve better coordination. Secretary of the Air Force Edward C. Aldridge stated that "some form of space command" is critical for the operation of space services. In the U.S. Congress, Senator John Warner called for space options for fighting a nuclear war both offensively and defensively. Representative Ken Kramer presented Resolution 5130 that proposed renaming the USAF as the "Aerospace Force."[260] This resolution also called for the creation of a separate space command. However, many events took place before the actual creation of an Air Force Space Command came into fruition.

First, the General Accounting Office released a report that questioned how the DoD was managing space systems. The report stated that a single manager should be in charge of military space endeavors and endorsed the CSOC to serve as the primary entity for it. This report also recommended withholding funding until the DoD presented a plan that was financially feasible.[261] To explore a proposal from AFSC Commander General Robert

---

[255]Spires, 198.

[256] Hall and Neufeld, 141.

[257] Hall and Neufeld, 142.

[258]Spires, 198.

[259]Spires, 201.

[260]Spires, 202.

[261]Boehm, 27.

Marsh on space management practices within the Air Force, USAF Chief of Staff General Lew Allen ordered studies to be done for initiatives to improve them. On April 15, 1982, General O'Malley flew to Colorado and approved a space command organizational flow chart.[262] Shortly after, the Air Force Space Command (AFSPC) was officially activated on September 1, 1982. On the bottom of the initial announcement for this command, General O'Malley penciled in that "it is the Air Force belief that the Air Force Space Command will develop quickly into a unified command with the Navy, Army, and Marine Corps."[263] NORAD Commander in Chief, General James V. Hartinger said that this command would have the "operational pull to go along with the technology push"[264] to really enhance space developments. Space Command hit that ground running and took responsibility for the development of MILSTAR, a satellite system that would provide reliable satellite communications to users worldwide for years to come,[265] and operational control of the emerging Global Positioning System.[266] At the same time, more unification was taking place with the Space Technology Center combining the tasks of three AFSC laboratories conducting research on geophysics, rocket propulsion, and space weapons.[267] Three years after the establishment of AFSPC, the vision of an integrated space command became reality with the creation of the United States Space Command.

---

[262]Hall and Neufeld, 142.

[263]Hall and Neufeld, 143.

[264]Spires, 205.

[265] Matthew Coleman-Foster, "Milstar Program Reaches 25 Year Milestone," U.S. Strategic Command, last modified February 6, 2019, https://www.stratcom.mil/Media/News/News-Article-View/Article/1760172/milstar-program-reaches-25-year-milestone/.

[266]Roeder.

[267]Spires, 205.

3. AFSPC Grooming for Primary Agent in Military Space Missions: 1986–2002

The year 1986 was a "dark age"[268] for launch capabilities within the USAF and NASA.[269] It started out hopeful with the developing interest of undertaking polar shuttle missions from the NRO as well as the USAF.[270] The USAF even created a launch pad in Vandenberg, California, for that very possibility.[271] The STS-62A Discovery mission was scheduled to use the Vandenberg launch pad to put the satellite *Teal Ruby*[272] into orbit but was cancelled after the Challenger explosion in January of 1986.[273] The loss of Challenger led to the USAF's termination of plans to use Vandenberg AFB for shuttle launches.[274] The decision to stop launching DoD payloads with the shuttle and to create the new Expendable Launch Vehicle Program (EELV) were results of the failures of 1986.[275] The USAF would not revisit the opportunity to launch from Vandenberg again until October of 1990, when the AFSPC assumed responsibility for space launch and reacquired Vandenberg, as well as Patrick AFB in Florida.[276] Though launch operations played a huge role in space tasks throughout the Cold War, understanding how to exploit space-based resources did as well. In the fall of 1992, however, a Blue Ribbon study discovered deficits in the USAF's ability to use space capabilities.[277] The study recommended the

---

[268] Hall and Neufeld, 147.

[269] Peebles, 29.

[270] Elizabeth Howell, "Classified Shuttle Missions," *Spaceflight*, last modified October 26, 2016, https://www.space.com/34522-secret-shuttle-missions.html.

[271] Howell.

[272] Howell.

[273] Peebles, 29.

[274] Hall and Neufeld, 147.

[275] Hall and Neufeld, 147.

[276] Roeder.

[277] Air Force Space Command "30th Anniversary Milestone: AFSPC Space Warfare Center (SWC) activated at Falcon AFB, Colorado," last modified November 7, 2012. https://www.afspc.af.mil/News/Article-Display/Article/249281/30th-anniversary-milestone-afspc-space-warfare-center-swc-activated-at-falcon-a/.

establishment of a Space Warfare Command, later named the Space Innovation and Development Center, which on November 1, 1993 was established and stood-up.[278]

With funding for USAF space assets at the forefront of leaderships' minds, the 2001 Space Commission made important organizational changes that resulted in the AFSPC moving acquisition internal to the command, the only USAF unit to have this structure.[279] This Commission spoke of the need for more cooperation between the USAF and the NRO and it recommended that they return to their original relationship.[280] Nevertheless, the call for this recommendation was not implemented. The Commission's recommendation that AFSPC become a separate four-star level USAF command was adopted in April of 2002. As a part of the initiative that boosted the U.S. military into a "21st century fighting force" and created U.S. NORTHCOM,[281] U.S. Space Command was dissolved and its mission was picked up by U.S. Strategic Command.[282] As a result, AFSPC once again became the only devoted command for military space undertakings.[283]

D.   USAF EXPANSION AND CURRENT MISSIONS: 2005-PRESENT

1.   Mission Area Cyberspace

The concept of having a domain presence in every area of warfighting is continually growing and in December of 2005, the USAF included the cyberspace

---

[278] Air Force Space Command, "30th Anniversary Milestone."

[279] "Air Force Space Command History," Air Force Space Command, accessed August 27, 2019, https://www.afspc.af.mil/About-Us/AFSPC-History/.

[280] George W. Bradley III, "A Brief History of The Air Force in Space," *High Frontier*, 2004,8, https://www.afspc.af.mil/Portals/3/documents/HF/AFD-070622-056.pdf

[281] Air Force Space Command, "AFSPC Milestone: AFSPC Becomes a Separate Four-Star Command," last modified August 13, 2012, https://www.afspc.af.mil/News/Article-Display/Article/249377/afspc-milestone-afspc-becomes-a-separate-four-star-command/.

[282] Air Force Space Command, "AFSPC Milestone."

[283] Roeder.

domain.[284] The service found it in the interest of the U.S. to expand its mission area to include "flying and fighting in Air, Space and Cyberspace."[285] The emergence of the cyber domain affected operations at AFSPC, as the responsibly for operations fell on the command with the Twenty-Fourth Air Force (24AF) as a subordinate established in 2009.[286] The cyber mission was not transferred to the Air Combat Command until July of 2018. The move helped AFSPC focus deeper on "gaining and maintaining space superiority."[287]

2.   Current Missions

With more than 26,000 USAF personnel operating within AFSPC alone, the USAF has a large space presence. There are many current space missions the USAF is conducting and supporting. The most visible system the USAF owns is the satellite-based navigation system, the Global Positioning System (GPS).[288] Another satellite constellation operated by the USAF is the Geosynchronous Space Situational Awareness Program (GSSAP).[289] This satellite program currently supports USSPACECOM surveillance operations in space. The USAF's secret X-37B space plane continues to break records in spaceflight duration.[290] The Space Fence, originally designed by the Naval Research Lab and transferred to the USAF in 2004 for operational authority, but then shut down in 2013, is being replaced by an S-band radar that has the ability to track a greater amount of smaller

---

[284] U.S Air Force, "Cyberspace as A Domain in Which the Air Force Flies and Fights," last modified November 2, 2006, https://www.af.mil/About-Us/Speeches-Archive/Display/Article/143968/cyberspace-as-a-domain-in-which-the-air-force-flies-and-fights/

[285] "Air Force Space Command History."

[286] "Air Force Space Command History."

[287] "Air Force Space Command History."

[288] "Air Force Space Command History."

[289] "Geosynchronous Space Situational Awareness Program Fact Sheet," AFSC, March 22, 2017, https://www.afspc.af.mil/About-Us/Fact-Sheets/Display/Article/730802/geosynchronous-space-situational-awareness-program/.

[290] Mike Wall, "X-37B Military Space Plane Breaks Record on Latest Mystery Mission," Space.com, last modified August 26, 2019, https://www.space.com/x-37b-military-space-plane-otv5-duration-record.html.

objects than prior systems in hopes of averting collisions in space.[291] The design review of the Space Fence has been completed on Kwajalein Atoll in the Marshall Islands and the technology is currently being produced that will bring the system to an operational status.[292] The USAF is currently testing on a scaled down integration test bed version in Moorestown, New Jersey.[293] The EELV, the future of guaranteed access to space, has two launch vehicles, the Atlas V and Delta IV, which are both operated by the USAF.[294] This program is managed by the Launch and Range Systems Directorate of the Space and Missile Systems Center, a component of AFSPC, located on Los Angeles Air Force Base.[295] The USAF Satellite Control Network (AFSCN), a network of satellite- tracking stations, is also a major system that is operated by the Air Force. This network provides communication uplink and downlink, as well as other tracking data to the warfighter.[296] Innovative space programs are also constantly being researched and developed within the USAF to maintain space awareness and the advancements that come along with it.[297] The USAF development and integration with other entities is what makes these missions listed and many others successful.

---

[291]Mike Gruss, "Haney: U.S. Partners to Have Indirect Access to Space Fence Data," *Space News*, last modified November 21, 2014, https://spacenews.com/42619haney-us-partners-to-have-indirect-access-to-space-fence-data/.

[292] "Space Fence," Lockheed Martin, Date Accessed November 15, 2019, https://www.lockheedmartin.com/en-us/products/space-fence.html.

[293]"Lockheed Martin, "Space Fence."

[294]"Evolved Expendable Launch Vehicle (EELV)," Vandenberg Air force Base, last modified August 4, 2017,

https://www.vandenberg.af.mil/About-Us/Fact-Sheets/Display/Article/1266632/evolved-expendable-launch-vehicle-eelv/

[295]"Evolved Expendable Launch Vehicle (EELV)."

[296]"Air Force Space Command History."

[297]"Air Force Space Command History."

E. USAF PERSPECTIVES ON POTENTIAL SPACE FORCE/CORPS INCORPORATION

1. USSPACECOM Reestablishment and USAF Integration

With USSPACECOM now reestablished, its commander, General John Raymond, is also dual hatted as the commander of AFSPC. One of the two direct subordinate organizations under USSPACECOM is also being run by Air Force leadership, the Combined Force Space Component Command (CFSCC).[298] The CFSCC's main goal is to deliver combat relevant space capabilities on all levels (strategic, operational, and tactical). Major General Stephen N. Whiting is the CFSCC commander.[299] CFSCC works closely with operational centers such as the Combined Space Operations Center (CSpOC) at Vandenberg AFB. The other subordinate command is Joint Task Force Space Defense (JTF-SD), responsible for "space superiority operations,"[300] and is headed by Brig. General Thomas James.[301] On April 3, 2019, General Raymond fully endorsed the Trump administration's proposal in front of the House Armed Services Committee (HASC) for a stand up of a Space Force under the USAF. General Raymond noted the importance and value of the Space Force four-star general "coming to work every day focused on the space domain."[302]

---

[298]"United States Space Command Fact Sheet," Department of Defense, August 29, 2019, https://www.spacecom.mil/About/Fact-Sheets-Editor/Article/1948216/united-states-space-command-fact-sheet/.

[299] Cody Chiles, "Combined Force Space Component Command Established At Vandenberg AFB," Schriever Air Force Base, last modified August 30, 2019, https://www.schriever.af.mil/News/Article-Display/Article/1948650/combined-force-space-component-command-established-at-vandenberg-afb/.

[300] Sandra Erwin, "Army General to Run One of USSPACECOM's Subordinate Commands," *Space News*, last modified September 1, 2019, https://spacenews.com/army-general-to-run-one-of-usspacecoms-subordinate-commands/.

[301]Erwin.

[302] Sandra Erwin, "Raymond Endorses Trump's Space Force Proposal," *Space News*, last modified April 3, 2019, https://spacenews.com/raymond-endorses-trumps-space-force-proposal/.

2. USAF Leadership on Both Sides of the Space Force/ Corps Debate

As discussed previously, the idea of a Space Force/Corps is not new. The main rationale has been to improve any budgetary efficiencies and to elevate the priority of space. The Government Accountability Office (GAO) released a comprehensive report in 2017 that described efforts to fix space acquisitions in the past as ineffective.[303] When it comes to air superiority or space dominance, the USAF has a track record of choosing air every time, and rightfully so, as that is the main mission.[304] The USAF has so many operations to juggle, including space, making it—according to some analysts-- impossible for the service to properly sustain all the competing needs.[305] Some senior officials, such as Major General James B. Armor, Jr., have agreed with these charges, but also make clear that "Air Force leadership propelled space power further than it often receives credit for today."[306]

This is why Gen. Armor suggested back in the early 2000s that the way to fix this problem is to form a Space Corps under the existing USAF organization, but with a separate budget.[307] He argued that because there were few space professionals and a small space budget that it was not feasible to create an entirely separate force.[308] The previous statement has also been an argument against the idea of a Space Force or Corps today. Following calls for a Space Corps in the House version of the 2018 National Defense Authorization Act, then-commander of USSTRATCOM, General John Hyten stated that the U.S. did not need a separate space service at the present moment due to space

---

[303] Phillip Swarts, "Air Force Lays out Its Case for Keeping Space Operations," *Space News*, last modified May 19, 2017, https://spacenews.com/air-force-lays-out-its-case-for-keeping-space-operations/.

[304] James B. Armor, Jr., "The Air Force's Other Blind Spot," last modified September 15, 2008, http://www.thespacereview.com/article/1213/1.

[305]Armor.

[306] James B. Armor, Jr., "Viewpoint: It Is Time to Create A United States Air Force Space Corps," *Astropolitics 5*, (2007): 278, DOI: 10.1080/14777620701580851.

[307] Armor, 274.

[308] Armor, 282.

representing a "very small force."[309] He did state, however, that he appreciated President Trump's description of space as a warfighting domain, which directs attention to the growing importance of space in the years to come.[310] Thus, General Hyten saw an eventual need for the creation of a Space Corps or Space Force, just not at the current moment.[311] Earlier, during testimony before the House Armed Services Committee in 2017, Chief of Staff General David Goldfein stated that creation of a Space Corps would actually separate space, when really space should be integrated within everything.[312] Generals Goldfein and Hyten fall into the same school of thought, which argues that the discussion should continue and remain important, but that a separate organizational restructure right now would not be smart. The reason Congressman Mike Rogers initially came up with the Space Corps concept was his belief that "the Air Force was prioritizing its fighter jets over space, and a dedicated service was needed to stay ahead of China and Russia in what many see as the next frontier of warfare."[313] But former Secretary of the Air Force Deborah Lee James stated—following President Trump's call for a U.S. Space Force in June 2018—that the creation of a separate service was not the right answer to solving this problem. She argued that personnel working on USAF space missions "will get totally lost in the bureaucracy of a Space Force."[314] She suggested instead that a new unified combatant command be stood up focused solely on space, but to keep the existing USAF space organizational structure as it stands.[315]

---

[309] Lauren C. Williams, "Now's Not the Time for a Space Force, STRATCOM Leader says," The Business of Federal Technology, last modified March 21, 2018, https://fcw.com/articles/2018/03/21/space-stratcom-sasc-hyten.aspx.

[310] Williams.

[311] Williams.

[312] Jeremy Herb,"House Passes Defense Bill That Would Create 'Space Corps',"CNN Politics, last modified July 14, 2017, https://www.cnn.com/2017/07/14/politics/house-passes-defense-bill-space-corps/index.html'.

[313] Herb.

[314] Oriana Pawlyk, "Former SecAF: Airmen Will Get Lost In 'Space Force'," DoD Buzz, last modified July 30, 2018, https://www.military.com/dodbuzz/2018/07/30/former-secaf-airmen-will-get-lost-space-force.html.

[315] Pawlyk.

The trend to keep USAF space operations within the USAF has been a clear inclination amongst some USAF leadership. An example of that leadership was the former Secretary of the Air Force Heather Wilson. In attempt to keep responsibility for space within the USAF, she endorsed in 2017 the creation of a new deputy chief of staff, the A-11 position on the Air Staff with the intent to "integrate, normalize, and elevate space within the service."[316] This new position also has a goal of developing more space professionals, which in return also heightens the visibility of space's significance.

Wilson has argued in the past that taking away space from the USAF would negatively affect space integration within the other services, but her attitude and the attitudes of many USAF leaders have since changed.[317] Just a few months after President Donald Trump announced his plan for forming a Space Force, Wilson announced that she was in "complete alignment" and said the president's plans had foresight.[318] Even Goldfein's approach softened after the announcement to a more supportive stance. However, Wilson's so-called alignment was not received well and she resigned in May of 2019. Her replacement, Barbara Barrett, says that she "is in full support of the establishment of the U.S. Space Force."[319] USAF leadership has used all the right words when describing this new Space Force, however, what they are really describing is a Space Corps, a body that will remain under the Air Force and functionality allow it to keep control of space. Despite past Congressional complaints, the USAF has managed to convince both the House and Senate during negotiations this year on the 2020 National Defense Authorization Act that whatever is formed—a corps (House version) or a force (Senate

---

[316] Phillip Swarts, "Air Force Sec. Wilson Makes New Space Leadership Position Official," *Space News*, last modified June 16, 2017, https://spacenews.com/air-force-sec-wilson-makes-new-space-leadership-position-official/.

[317] Valerie Insinna, "Top U.S. Air Force Official Is Now on Board With Trump's Space Force plan," *Defense News*, last modified September 5, 2018, https://www.defensenews.com/smr/defense-news-conference/2018/09/05/the-top-air-force-official-is-now-onboard-with-trumps-space-force-plan/.

[318] Insinna.

[319] Sandra Erwin, "Air Force nominee Barret Calls for Assertive U.S. Posture on Space, Says Space force is A 'Key Imperative'," *Space News*, last modiefied September 12, 2019, https://spacenews.com/air-force-nominee-barrett-calls-for-assertive-u-s-posture-on-space/

version)--it will reside within the USAF. The USAF use of the term Space Force appeases President Donald Trump at the same time.

F.   CONCLUSION

The USAF has been active from the very beginning in regard to space. However, more often than not, it has not been smart with its funding of space due to its task to manage multiple missions. USAF leadership has historically failed to sufficiently provide for the space missions assigned to it. Though the USAF is leading in space technology and innovations, it is lacking in effectiveness and efficiency, which has resulted in duplication of efforts with other services. The USAF itself has been in search of a consolidated space entity since the early 1960s and somewhat achieved this goal with creation of the AFSPC. With the rebirth of USSPACECOM and with the USAF's sizable presence in staffing and leadership roles, along with being involved with some civilian launches, the USAF's space mission is focused on maintaining space superiority, but critics argue that there needs to be a better management of resources. How exactly the USAF will support the stand-up of the coming Space Force/ Corps is still being debated.

V.   THE U.S. SPACE FORCE / CORPS DEBATE

A.   INTRODUCTION

The previous chapters have provided a historical outline that now helps shape the current Space Force debate. This chapter will introduce President Donald Trump's 2018 call for a Military Space Force and briefly summarize the Space Policy Directive-4. It will then discuss and suggest current and potential future organizational structures by discussing what the DoD and members of Congress have already projected in relation to the United States Space Command (USSPACECOM) and the Space Development Agency (SDA). After, the chapter will revisit previously stated hypotheses, and deliberate them, while discussing possible future policy actions. This chapter will then suggest further areas for study and culminate with concluding overall remarks on the fate of other space units and why the AFSPC should be renamed the Space Corps and remain under the USAF.

B.   CURRENT SPACE FORCE / CORPS DEBATE

On June 18, 2018, President Donald J. Trump said that "When it comes to defending America, it is not merely enough to have an American presence in space, we must have American dominance in space…I'm hereby directing the Department of Defense and Pentagon to immediately begin the process necessary to establish a Space Force as the sixth branch of the Armed Forces. That's a big statement."[320]

A big statement it was, as it left many mystified on what exactly a Space Force would look like within the Department of Defense (DoD). President Trump stated that this Space Force would be "separate but equal from the USAF," suggesting that he wanted to see a sixth branch of the military outside of the USAF initially. As stated earlier, the last time a military service was separated was in 1947. When the White House submitted a legislative proposal called the Space Policy Directive-4, it called for the DoD to officially

---

[320] White House, *Remarks by President Trump at a Meeting with the National Space Council and Signing of Space Policy Directive-3* (Washington, DC: White House, 2018), https://www.whitehouse.gov/briefings-statements/remarks-president-trump-meeting-national-space-council-signing-space-policy-directive-3/.

construct a military service for space. This Directive was signed by President Trump in February of 2019, and the text outlined what the Department of the Space Force would be. It stated that "it refers to a new branch of the United States Armed Forces to be initially placed by statute within the Department of the Air Force."[321] But the directive essentially described a Space Corps inside the USAF, and it gave many a sigh of relief, as the administration had not decided to jump straight into creating a separate armed service apart from the USAF. The directive also discussed the consolidation of existing forces and authorities to minimize duplication, a problem that has haunted the DoD in the past. While it did not propose combining the military's space forces, it annotated that this did not include "the National Aeronautics and Space Administration, the National Oceanic and Atmospheric Administration, the National Reconnaissance Office, or other non-military space organizations or missions of the United States Government."[322] It also discusses the "Associated Elements" that will be attached to the Space Force, and stated that U.S. Space Command would be reinstated. The Space Force Strategic Overview, released in February of 2019, states that the goal of establishing USSPACECOM is to "bring full-time operational focus to securing the space domain."[323] Figure 1 illustrates the original Space Force paradigm created by the DoD.[324] The red writing annotates the changes that are proposed to be made within the Department of the Air Force, the Joint Chiefs, and the unified combatant commands. In this Directive, an active duty 4-star would serve as Chief of Staff of the Space Force under the Joint Chief of Staffs and a civilian would serve as Secretary for Space.[325]

---

[321] Marcia Smith, "Text of Space Policy Directive-4 (SPD-4): Establishing a U.S. Space Force," Space Policy Online, February 19, 2019, (Presidential Memorandum. White House, 2019) https://spacepolicyonline.com/news/text-of-space-policy-directive-4-spd-4-establishing-a-u-s-space-force/.

[322] Smith.

[323] Department of Defense, *United States Space Force* (Washington, DC: Department of Defense, 2019), 7, https://media.defense.gov/2019/Mar/01/2002095012/-1/-1/1/UNITED-STATES-SPACE-FORCE-STRATEGIC-OVERVIEW.PDF.

[324] Department of Defense, *United States Space Force,* 6.

[325] Department of Defense, *United States Space Force,* 1.

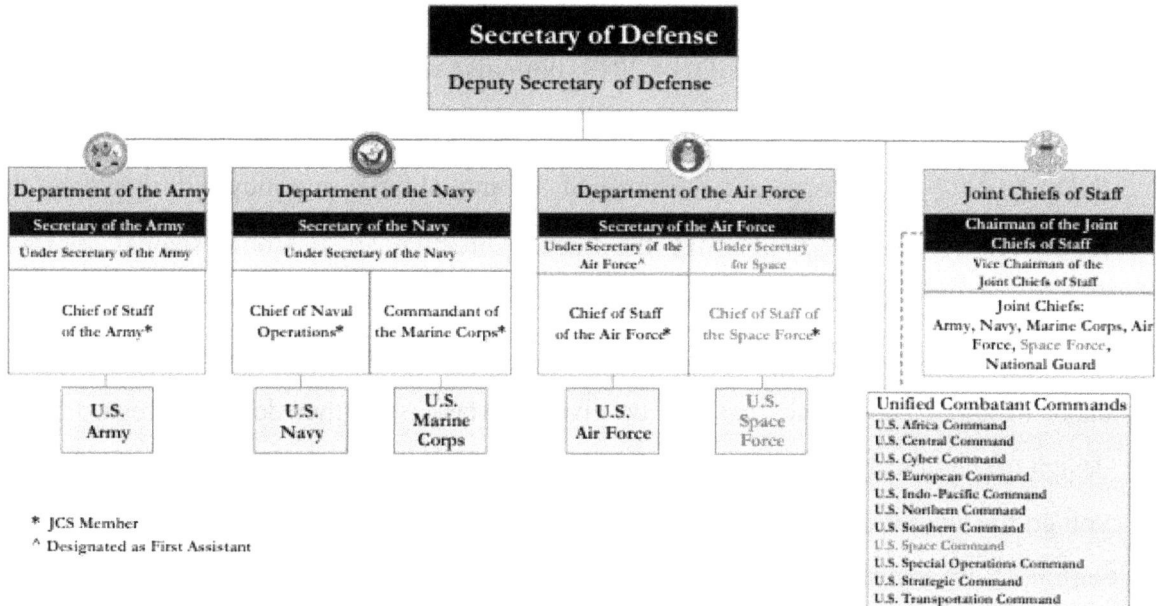

Figure 1.   Space Directive-4 U.S. Space Force structure within the
DoD[326]

1.   SASC/HASC Approvals of Service within the NDAA

President Trump's approval of the Space Directive-4 permitted the White House to submit to Congress a request to stand up this U.S. Space Force. Since then, there have been many meetings within the Senate Armed Services Committee (SASC) and the House Armed Services Committee (HASC), as they are the subject matter experts to Congress in regard to a way forward for space efforts.[327] Since these two dedicated committees are a part of Congress, their legislative proposals are essential for an ultimate decision. HASC's Chairman Representative Adam Smith has his reservations in regard to this new service. However, his fellow colleague and the HASC leader of the Strategic Forces Subcommittee, Representative Jim Cooper, has been an advocate of the concept.[328] While members of the

---

[326] Source: Department of Defense, *United States Space Force,* 6.

[327] Kaitlyn Johnson, "Space Force or Space Corps?," Center for Strategic and International Studies, last modified June 27, 2019, https://www.csis.org/analysis/space-force-or-space-corps.

[328] Valerie Insinna, "Senate Authorizers Approve Space Force but Switch Up its Organizational Structure," last modified May 23, 2019, https://www.defensenews.com/space/2019/05/23/senate-authorizers-approve-space-force-but-switch-up-its-organizational-structure/.

HASC and SASC have all expressed views for and against the creation of the Space Force, in May (SASC) and June (HASC) of 2019, both also released their respective versions of "military space reorganization"[329] within the National Defense Authorization Act (NDAA).[330] Since Congress in is charge of the annual defense policy bill, they have the final decision whether to approve the Space Directive.[331] If Congressional approval happens, the U.S. Space Force could be stood up within 90 days, said the previously acting Secretary of the Air Force, Matt Donovan, who confirmed the service already has a space cadre of about 200 people that will be given the task of further details in regard to the design of the branch. He stated that "these 200 people will then be given the task of: 'Now you go flesh it out. Put the meat on the bones of how we're going to design a Space Force, with not only the staff, but the entire organization.' How are we going to be organized to provide the best support to the war fighter?"[332]

SASC Chairman Jim Inhofe stated that "we know space is a warfighting domain, so we are setting up the U.S. Space Force with the Air Force. Our strategy will set the Space Force up for success now and in the future by minimizing bureaucracy."[333] These specific versions (H.R. 2500 and the S.1790) from the HASC and SASC are definitely a step forward, but each comes with specific requirements. Thus, similarities are present within both competing versions of the FY2020 NDAA, the differences are very clear as well. Though the explanations of specific changes are to follow, Figures 2 depicts key points by lining up the three versions (DoD, SASC, and HASC) as to what this new space entity might look like in a chart format. To note, the first asterisk within the chart by the "Yes" under SASC FY20 NDAA represents the SASC not overtly stating in the markup that a new service is being created within the USAF. The asterisks by the "four-star" also

---

[329] Johnson.

[330] Johnson.

[331] Insinna.

[332] Valerie Insinna, "Documents Reveal How the Space Force Would Launch in 90 Days," *Defense News*, last modified September 16, 2019, https://www.defensenews.com/digital-show-dailies/2019/09/16/documents-reveal-how-the-space-force-would-launch-in-90-days/.

[333] Insinna. "Senate Authorizers."

in the SASC column is a nuance that for the initial year, the Commander of U.S. Space Command will be dual hatted and also function as the U.S. Space Force Commander. [334]

	DOD LEGISLATIVE PROPOSAL	SASC FY20 NDAA MARKUP	HASC FY20 NDAA AMENDMENT
Creates a New Military Department	No	No	No
Creates a New Service within the Air Force	Yes	Yes*	Yes
Creates a New Civilian Space Position in the Air Force	Yes, Under Secretary of the Air Force for Space	No	No
Senior OSD Civilian Position	No	Elevates Deputy Assistant Secretary of Defense (DASD) for Space to be Assistant Secretary of Defense (ASD) for Space Policy	No
Military Leadership	Chief of Staff, four-star Vice Chief of Staff, four-star	Commander, four-star** Vice-Commander, four-star	Commandant, four-star
Representation on the Joint Chiefs of Staff	Yes	Yes	Yes
New Civil Service Positions	Yes and gives greater flexibility for recruiting, hiring, and pay for civilians in the Space Force	No, transfers existing billets	No, transfers existing billets
New Military Positions	Yes, allows the Secretary of Defense to authorize new military positions for the Space Force	No, transfers existing billets	No, transfers existing billets
Includes NRO and other intelligence Agencies	No	No	No
Includes Army and Navy space components	Yes	No	Not immediately but requires DoD to submit a report to Congress advising necessity
Includes National Guard and Reserve Components	Includes associated reservists, but left out the National Guard	Yes	Unclear
Mentions SPACECOM	Yes	Yes	Yes
SPACECOM LEADERSHIP	Commander, four-star	Commander, four-star**	Not addressed
Mentions the SDA	No	Yes	Yes
Estimated Budget	Gives DoD authority to transfer funds and establish a headquarters for the Space Force	Intended to be budget neutral	Requires a report on the estimated funding requirements to establish and operate the Space Corps through the FYDP (2021-2025)
New Service Transition Timeline	5 years from enactment	Not addressed	December 30, 2023

Figure 2.   DoD, SASC, and HASC assessments of new service [335]

---

[334] Johnson.

[335] Source: Johnson.

65

2. Noted SASC/HASC Amendments to the Space Policy Directive-4

The most noticeable differences between H.R. 2500 and S. 1790 that diverge from the White House's initial proposal include lack of a new civilian position for space within the USAF structure, and for a direct report to the Secretary of the Air Force. They both have the methodology of a corps structure, within the Department of the Air Force.[336] Though all write-ups endorse the addition of a four-star general leading the new space service, the SASC says that the first year after enactment of the service, the role of the Commander of AFSPC will be renamed as Commander of the United States Space Force (USSF) and that position will report to the Secretary of the Air Force (SecAF) by way of the Air Force Chief of Staff. After one year, the position will report directly to the SecAF. Also throughout the first year, the Commander of USSF will be invited to meetings with the Joint Chiefs of Staff for discussion of space issues; afterward, the position will become a permanent member of the Joint Chiefs of Staff.[337] H.R. 2500 and S. 1790 also both state there will be no new military and civilian positions, and that all existing positions will transfer.[338]There are differences between the two committees' versions as of what the proposed new service will be called. The SASC would use President Trump's initial name of the U.S. Space Force. The HASC has officially entitled the new service the U.S. Space Corps. Though this is a very visible difference, it isn't the most important one, as the titles still reference the same thing, a structure in which the Force/Corps would be under the Department of the Air Force initially.[339] H.R. 2500 proposes a transition period of January 1, 2021, through December 30, 2023.[340] S. 1790 does not address the timeline but does state that a report to Congress must be provided listing out the structure and cost by January

---

[336] Johnson.

[337] Insinna, "Senate Authorizers."

[338] *Military Space Reform: FY2020 NDAA Legislative Proposals,* CRS Report No. 7–5700 (Washington, DC: Congressional Research Service, 2019), 2, https://fas.org/sgp/crs/natsec/IF11326.pdf.

[339] Military Space Reform, 2.

[340] Military Space Reform, 2.

of 2021.[341] The final differences will be in regard to USSPACECOM and the SDA. With both stood up currently, I will now discuss them more in further detail.

3.   U.S. Space Command Reestablishment Concerns

The newest of the eleven unified commands is now stood up and General Raymond has been authorized to build a staff of roughly 300, but USSPACECOM is far from being fully operational.[342] Criticisms about the order of precedence must be addressed in regard to the creation of this entity. Though USSPACECOM was officially stood up on August 29, 2019, since creating a new unified combatant command does not require the approval of congress,[343] it is important to discuss possible problems that could affect the future. Commentator Brian Weeden argues that USSPACECOM should have been established after the Space Force (under the USAF) was officially stood up.[344] Weeden's key argument is that instead of separating space from other combatant commands, it should be integrated by multiplying space domain experts within existing combatant commands. Service members that have filled space billets within other combatant commands have reported a shortage of space experts, as many space billets have not been filled. This may be a problem within USSPACECOM down the line in regard to the proper manning of the combatant command. In addition, Weeden argues that a "U.S. Space Command that is culturally biased towards focusing on space as a separate warfighting domain, as opposed to space being part of a broader military picture that incorporates terrestrial military operations, would make things worse."[345] With USSPACECOM being a combatant

---

[341] Military Space Reform, 2.

[342] Sandra Erwin, "Five Things to Know about the Space Command," *Space News*, last modified October 23, 2019, https://spacenews.com/five-things-to-know-about-u-s-space-command/.

[343] Marcia Smith, "Space Command Gets to Work While Congress Continues to Debate Space Force," *Space Policy Online* (September 2019), https://spacepolicyonline.com/news/space-command-gets-to-work-while-congress-continues-to-debate-space-force/.

[344] Brian Weeden, "Space Force is More Important than Space Command," Texas National Security Review, last modified July 8, 2019, https://warontherocks.com/2019/07/space-force-is-more-important-than-space-command/.

[345] Weeden.

command, General Hyten explains that "it will execute the space mission that fights our forces and wins our wars."[346] If the Space Force/Corps is officially stood up, it will organize, train, and equip for USSPACECOM. With General Hyten breaking down the difference between the two, Weeden has a point. The foundation seems backward. How do you fight a war that you have not organized for, you have not trained for, and you have yet to equip for? The point of consolidation was to reduce confusion. However, this order is unclear. This argument will be expounded upon later on in this chapter to discuss why AFSPC has already done the job of the Space Force/Corps in regard to this organize, train, and equip portion. SDA, established a little more than six months before USSPACECOM, on March 12, 2019, arguably helps with this as well, as the architecture is meant to "unify and integrate efforts across the department"[347] in regard to space threats.

4.   Establishment of the SDA

Leadership within the Pentagon originally saw the need for the SDA due to the vulnerability of the military's space architecture to possible attack with its "small constellations of large, exquisite satellites."[348] Thus, according to Director Derek Tournear, the SDA's vision consists of three words: "resiliency via numbers."[349] The goal is to have multiple small satellites all operating in a network so, if one is lost, it will not be debilitating, which is the risk with the present space architecture. Building resiliency to replace our current reliance on a "constellation of juicy targets," as Tournear puts it, is the SDA's goal.[350] To do this, the SDA will rely on already developed and developing

---

[346] United States Strategic Command, "Difference between USSPACECOM and Space Force?" April 12, 2019, https://www.facebook.com/usstrategiccommand/videos/380744895845035/.

[347] Sandra Erwin, "Shanahan Officially Establishes the Space Development Agency," *Space News*, last modified March 13, 2019, https://spacenews.com/shanahan-officially-establishes-the-space-development-agency/.

[348] Nathan Strout, "What Will the Space Development Agency Really Do?"*C4ISRNET*, last modified July 24, 2019, https://www.c4isrnet.com/battlefield-tech/space/2019/07/24/what-will-the-space-development-agency-really-do/.

[349] Strout.

[350] Strout.

capabilities, instead of starting from scratch. This approach is said to reduce redundancy in efforts, which has been arguably a reoccurring problem throughout the space age.

C.   DISCUSSION OF HYPOTHESES

So, what is the path that actually supports the warfighter most efficiently? In the beginning of this thesis, three hypotheses were delineated. To reiterate, the first hypothesis stated that there should be an independent space service, but after years as a Corps-like structure under the USAF (SASC or HASC dependent). The second stated that there should be a Space Corps that would remain under the USAF, much like the structure of the U.S. Marine Corps under the Navy. The third hypothesis was that the Space Directive-4 should be rejected, and the services should remain on the current course, or adopt other reforms. Also, within all of the hypotheses remains the question of what should be done with other DoD space components if these actions are taken.

Is a separate military space organization necessary? With the return of USSPACECOM, and the establishment of SDA, the DoD has proven they are serious about space and ongoing reforms. Strictly from an acquisitions standpoint, the GAO has proven that actions need to be taken in regard to a historical track record of the USAF failing in regard to inadequate funding of space, and the development of its assets. These actions start in creating a place for space funding to go. From a man, train, and equip aspect, the DoD needs a consolidated organization to which it can turn to for all things space. This structure does not separate space in the least, but instead, helps integrate it more as there will be a clear contact. These reforms, however, are not enough to stabilize and localize space efforts. The C2 aspect must be addressed, and space must be given an official home within the DoD. Though there is now a combatant command for space, where space missions will be executed from, there must be a dedicated manning, training, and equipping of forces for USSPACECOM. This is why the third hypothesis is not valid, and action must be taken in the form of centralization of all DoD space components.

However, it is not clear that a separate Space Force is the best answer to address this issue. Evidence shows that there is no need for an independent Space Force. Firstly, the argument that one cannot have a unified chain of command for space without having a Space Force is just not true. The February 2019 Space Force Strategic Overview stated that

> The Space Force will provide the preponderance of its forces to USSPACECOM and also provide forces to all CCMDs in order to integrate space capabilities and doctrine into all CCMD planning and operations. Combatant commands will communicate their space capability needs for both joint force enhancement and domain superiority to the Space Force and would coordinate with the Space Force to satisfy emerging and urgent operational needs in an expeditious manner.[351]

But this kind of reporting does not require a separate Space Force. As long as the DoD is clear on where different space assets are, space capability needs will be heard and reconciled. The creation of a completely separate Force is likely to be an unnecessary bureaucratic nightmare that may not pay off in the end.

The most feasible of these three hypothesis would be the second one, in which a Space Corps should be established under the USAF. This thesis recommends the HASC proposed name but sees benefit in using parts of both amendments. The SASC amendment states clearly that the Commander of USSPACECOM will also fill the role of the Commander of the Space Corps for the first year. This is beneficial because it will help leadership to be on the page from a C2 standpoint. However, the SASC markup also proposes an invite-only position within the Joint Chiefs of Staff for the first year for the Space Corps Commander. This does not flow toward the goal of creating this entity. The Commander needs to be fully integrated from day one so that a pattern can be started from the beginning that space is just as important as the other mission areas. Having an invite-upon request mentality will essentially downgrade the importance the Commander holds within the JCS. In going deeper on this hypothesis, it is recommended to rename AFSPC the Space Corps and augment from there. As stated, a huge portion of space-related DoD activities already exists in AFSPC. Even though the mission would essentially stay the same, this name change could spark more central funding for space by making it seen as more of a priority. With over 26,000 people within AFSPC, the command handles a wide

---

[351] Department of Defense, *United States Space Force,* 7.

spectrum of space. From satellite communications to secret projects like the X-37B space plane, AFSPC has a hand in most space-related projects. Lt. Gen. Bruce Wright (USAF, Ret.), states that "This new cadre of space warriors will draw directly from the proven performance and experience of Airmen and the structure of AFSPC."[352] So, why spend more government funding when the Space Corps can be AFSPC? Dedicating a budget to primarily space would end the talk of the USAF using space funding for other mission areas within the USAF and make the point that the funding is for space mission integration to the joint force at large. As stated, AFSPC already is the only command within the USAF to have acquisition internal to the command. The DoD should feed off of this structure and put all of the space budget within USSPACECOM, and then outsource specific missions and funding to other DoD components.

To ensure space integration happens with other space entities outside of the USAF, a Space Corps representative or team, dependent on how large the entity is, should be assigned to all joint space components outside of the Space Corps. For example, since NAVSOC was designated Satellite Control Authority by USSTRATCOM when it was in charge of space operations,[353] USSPACECOM should give that same authority and a Space Corps team should be integrated within it to guarantee the lines of communication and planning remain open and clear. It is also recommend for the Space Corps teams to be comprised of different services, so the USAF culture would not be carried over, and the Space Corps would feel and work from a joint perspective. The Space Corps teams would also fall in the chain of command as liaisons to the official Space Corps.

Given the knowledge and unique space missions within the other services, my recommendation would also be to keep their space cadres intact, but give cadre professionals the initial option to switch over if desired, or volunteer to liaison by joining a Space Corps team in a respective joint command, as long as mission can support. Moving all of the space cadre out of their respective services by force would not make sense,

---

[352] Bruce Wright, "Fighting and Winning in Space- Today and Tomorrow," Air Force Association, last modified August 14, 2019, https://www.afa.org/publications-news/media/president-perspective.

[353] DeJesus.

especially since the numbers show that it would not really even make a dent in impact since they are so small. With about 208 active-duty and reservists assigned to the U.S. Navy space cadre, and 389 space cadre professional officers on U.S. Army active-duty, in the Army National Guard, and in the U.S. Army Reserve, they will serve the space mission better by acting as space advocates in their respective services, which will further integrate the space mission. To address the need for manning of USSPACECOM and the SDA, as noted in President Trump's FY 2020 budget request, it is suggested to employ civilian workers, transferring those willing to move to active-duty, and utilizing the National Guard and Reserve components by offering additional billets. As new recruits join the armed forces, it makes sense to offer a direct track to space from each service as an initial option. This would address manning concerns as well. The space curriculum at the Naval Postgraduate should remained untouched, only that now the officers going through the program should have a dedicated space track that detailers are aware of, regardless if it is to be within their dedicated service, on a Space Corps team, as part of the AFSPC/Space Corps, or in USSPACECOM.

In regard to the U.S. Army and U.S. Navy space assets, such as MUOS, these assets should also remain intact to their respective services with the central space element and chain of command being AFSPC/Space Corps and USSPACECOM. Even the Acting Defense Secretary Patrick Shanahan stated that "the Navy and Army will retain control of their space assets." He says that moving these assets into a new organization is unnecessary right now because he states that the main focus "with new authority, is getting after the threat."[354] History has shown us that past restructuring has complicated things even more in regard to space. Critic Michael O'Hanlon states that if a Corps were created, it would be too small to even make sense or benefit from it as it requires its own chain of command. He argues that we "cannot create a new service for each partially neglected area of the

---

[354] Ben Werner, "Shanahan: Space Force Won't Take Over Navy, Army Space Assets," *USNI News*, last modified March 20, 2019, https://news.usni.org/2019/03/20/new-space-force-will-not-take-away-navy-space-assets.

armed forces."[355] He also reasons that all the tedious discussions and policy implementations would take the focus off what is really important: space. Instead, he suggests increasing space professionals in existing services. However, just because the Corps would be small does not mean it will not be necessary. In addition, O'Hanlon seems to downplay how important space is and how much it affects operations and daily integration. Current trends suggest that our space professionals will grow as the needs of space expand with emerging threats.

D.   CONCLUSIONS

1.   Future Study

Suggestions for future study are reliant on the decision Congress makes with the Space Policy Directive-4. After the decision, a deep dive into the funding side through GAO reports could be analyzed for better understanding of how space funding has been leveraged over the years to make sure it does not continue to be redirected to non-space missions. Another path moving forward is to focus on the unified combatant commands in more depth and describe how including USSPACECOM affects each of them and what integration with USSPACECOM looks like. Also, a focus on the intelligence community would be beneficial, as it is a big part of space. Research on how the decision Congress makes will or will not actually integrate these organizations such as the NRO is important to focus in on, and how their functions with the activities of the new Space Corps will matter to overall U.S. success in space. Also, if the Space Corps is stood up, a comparison of adversary space forces would be beneficial to understand how they structure their space efforts and how we differ to better understand how to defeat U.S. adversaries.

---

[355] Michael O'Hanlon, "The Space Force is a Misguided Idea. Congress Should Turn It Down," Brookings Institution, last modified April 20, 2019, https://www.brookings.edu/blog/order-from-chaos/2019/04/20/the-space-force-is-a-misguided-idea-congress-should-turn-it-down/.

2.   Overall Remarks / Conclusions

It is very likely that Congress will decide whether to support the newly proposed service in the near future. To understand how best to deliberate that choice, defining what the current problems are via an extensive historical analysis is key. Throughout the years, space policy has been complex, and, unfortunately, it continues to pose difficult problems. Throughout history, organizations and agencies have come and gone, different opinions have been heard or shut down, but the space debate has remained consistent. There has been a recurring theme to understand what each player actually contributes to the space game and to make sure these critical functions are preserved. Moving forward, viewing space as a warfighting domain, critical to the fight, will be essential for achieving the proper attention and funding space deserves.

LIST OF REFERENCES

Air Force Space Command. "30th Anniversary Milestone: AFSPC Space Warfare Center (SWC) Activated at Falcon AFB, Colorado." Last modified November 7, 2012. https://www.afspc.af.mil/News/Article-Display/Article/249281/30th-anniversary-milestone-afspc-space-warfare-center-swc-activated-at-falcon-a/.

———. "A Brief History of the U.S. Air Force." September 12, 2016. Accessed January 19, 2019. https://www.afspc.af.mil/News/Article-Display/Article/942428/a-brief-history-of-the-us-air-force/.

———. "Air Force Space Command History." Accessed August 27, 2019. https://www.afspc.af.mil/About-Us/AFSPC-History/.

———. "Geosynchronous Space Situational Awareness Program Fact Sheet." March 22, 2017. https://www.afspc.af.mil/About-Us/Fact-Sheets/Display/Article/730802/geosynchronous-space-situational-awareness-program/.

———. "AFSPC Milestone: AFSPC Becomes a Separate Four-Star Command." Last modified August 13, 2012. https://www.afspc.af.mil/News/Article-Display/Article/249377/afspc-milestone-afspc-becomes-a-separate-four-star-command/.

AMCOM. "The ABMA/AOMC Era, 1956–62." Accessed May 23, 2019. https://history.redstone.army.mil/ihist-1956-intro.html.

Armor, James B. Jr. "The Air Force's Other Blind Spot." Last modified September 15, 2008. http://www.thespacereview.com/article/1213/1.

———. "Viewpoint: It is Time to Create a United States Air Force Space Corps." *Astropolitics* 5, no. 3. (September-December 2007): 273–288. doi:10.1080/14777620701580851:280-281.

ASPDO. "FA40 Roster." August 21, 2019. file://comfort/jmjohnso2$/Downloads/FA40%20Roster-Ver%203%20(1).pdf

Atlas Obscura. "Peenemünde Army Research Center." Accessed September 16, 2019, https://www.atlasobscura.com/places/peenemunde-army-research-center-1.

Atomic Heritage Foundation. "Strategic Defense Initiative (SDI)." July 18, 2018. https://www.atomicheritage.org/history/strategic-defense-initiative-sdi.

Berkowitz, Bruce, and Michael Suk. *The National Reconnaissance Office at 50 Years: A Brief History*. Chantilly, VA: National Reconnaissance Office, Center for the Study of National Reconnaissance, 2015.

Bradley, George W. III. "A Brief History of the Air Force in Space." *High Frontier*. 2004.https://www.afspc.af.mil/Portals/3/documents/HF/AFD-070622-056.pdf

Brissett, Wilson. "The Space Corps Question." *Air Force Magazine*. October 2017. http://www.airforcemag.com/MagazineArchive/Pages/2017/October%202017/The-Space-Corps-Question.aspx.

Boehm, Joshua. "A History of United States National Security Space Management and Organization." Commission paper, Commission for Assessment of United States National Security Space Management and Organization, 2001. https://fas.org/spp/eprint/article03.html#ft78.

Burns, Robert. "Senators on Trump Space Force Plan: Not So Fast." Associated Press. Last modified April 11, 2019. https://www.apnews.com/5a6e1a08acb745a69976e3110b4789b4.

Brown, Brian B. "2018 Retrospective and Look Ahead for 2019." Official memorandum, Washington, DC: United States Navy,2019. https://www.mnp.navy.mil/documents/34084/0/Space+EOY+Summary+for+2018.pdf/290f72db-31e7-0646-914d-51ad2972e137.

Burrows, William E. *This New Ocean: The Story of the First Space Age*. New York, NY: Random House Publishing Group, 1998.

Campbell, Kevin. "The Army's Space Cadre," *High Frontier Journal* no. 1 (November 2007). https://documents.theblackvault.com/documents/space/AFD-071119-017.pdf.

Cavallaro, Emanuel. "Eyes on the Skies: NRL Researchers Tackle the Ever-Growing Problem of Orbital Debris," U.S. Naval Research Laboratory, https://www.nrl.navy.mil/space/eyes-skies-nrl-researchers-tackle-ever-growing-problem-orbital-debris.

Chiles, Cody. "Combined Force Space Component Command Established at Vandenberg AFB." Schriever Air Force Base. Last modified August 30, 2019. https://www.schriever.af.mil/News/Article-Display/Article/1948650/combined-force-space-component-command-established-at-vandenberg-afb/.

Christianson, Gale E. "Science Fiction Studies." Kepler's Somnium: Science Fiction and the Renaissance Scientist. Accessed March 3, 2019. https://www.depauw.edu/sfs/backissues/8/christianson8art.htm.

Clark, Colin, and Colin Clark. "Space Development Agency Opens; to Tackle Hypersonics from LEO." Breaking Defense. March 14, 2019. Accessed March 16, 2019. https://breakingdefense.com/2019/03/space-development-agency-opens-to-tackle-hypersonics-from-leo/.

Clark, Liesl. "Polynesia's Genius Navigators." NOVA. February 14, 2000. https://www.pbs.org/wgbh/nova/article/polynesia-genius-navigators/.

Coleman-Foster, Matthew. "Milstar Program Reaches 25 Year Milestone." U.S. Strategic Command. Last modified February 6, 2019. https://www.stratcom.mil/Media/News/News-Article-View/Article/1760172/milstar-program-reaches-25-year-milestone/.

Congressional Research Service. *Military Space Reform: FY2020 NDAA Legislative Proposals.* CRS Report No. 7–5700. Washington, DC. 2019.1-2.https://fas.org/sgp/crs/natsec/IF11326.pdf.

Cooper, Helene. "A Space Force? The Idea May Have Merit, Some Say." *The New York Times.* Last modified June 23, 2018. https://www.nytimes.com/2018/06/23/us/politics/trump-space-force-military.html.

Cozzens, Tracy. "U.S. Space Command Re-Established as 11th Unified Combatant Command." GPS World. Last modified August 29, 2019. https://www.gpsworld.com/u-s-space-command-re-established-as-11th-unified-combatant-command/.

Cutshaw, Jason B. "53rd Signal Battalion Leads the Army's Space Operations." U.S. Army. Last modified November 29, 2017. https://www.army.mil/article/197230/53rd_signal_battalion_leads_the_armys_space_operations.

———. "Celebrating the Army's Explorer 1 Legacy." U.S. Army. Last modified January 30, 2018. https://www.army.mil/article/199846/celebrating_the_armys_explorer_1_legacy.

DeBlois, Bruce M. "Space Sanctuary: A Viable National Strategy." *Air and Space Power Journal.* no. 1 (1998):41-57. https://cle.nps.edu/access/content/group/19ebe878-76c5-4974-b810-c080ee8102f6/Course%20Documents/Class%2012/DeBlois1998.pdf.

Deffree, Suzanne. "German Rocket Is 1st to Reach Space, October 3, 1942."EDN Network. Last modified October 3, 2018. https://www.edn.com/electronics-blogs/edn-moments/4397678/German-rocket-is-1st-to-reach-space--October-3--1942.

DeJesus, Adam. "Navy Space Cadre & Warfighters Putting Pieces Together." *CHIPS, The Department of the Navy Information Technology Magazine.* June 10, 2014. https://www.doncio.navy.mil/chips/ArticleDetails.aspx?ID=5186.

Department of Defense. *Development of Space Systems.* DoD Directive 5160. 32. Washington, DC: Department of Defense, 1970. https://apps.dtic.mil/dtic/tr/fulltext/u2/a272407.pdf

————. *Reconnaissance, Mapping and Geodetic Programs Reference.* DoD Directive 5160. 32. Washington, DC: Department of Defense,1961. https://www.nro.gov/Portals/65/documents/foia/declass/WS117L_Records/215.PDF.

————. *"United States Space Command Fact Sheet."* 2019. https://www.spacecom.mil/About/Fact-Sheets-Editor/Article/1948216/united-states-space-command-fact-sheet/.

————. *United States Space Force.* Washington, DC. 2019.1-10. https://media.defense.gov/2019/Mar/01/2002095012/-1/-1/1/UNITED-STATES-SPACE-FORCE-STRATEGIC-OVERVIEW.PDF.

Department of the Air Force. *U.S. Space Command Talking Points and RTQ.* STRATCOM: 12 April 2019.

Department of the Army. *Army Space Operations.* FM 3-14. Washington, DC: Department of the Army, 2014. https://fas.org/irp/doddir/army/fm3_14.pdf.

————. *Commissioned Officer Professional Development and Career Management.* Pamphlet 600–3.December 3, 2014. https://www.army.mil/e2/c/downloads/376665.pdf.

Department of the Navy, *From the Sea to the Stars: A Chronicle of the U.S. Navy's Space and Space Related Activities, 1944–2009.* Washington, DC: Department of the Navy, 2010. http://www.history.navy.mil/books/ space/FromTheSeaToTheStars-2010ed.pdf.

————. "MUOS Satellite Improves Communications for U.S. Forces on the Move." Navy Live. Accessed March 1, 2019. https://navylive.dodlive.mil/2013/07/18/muos-satellite-improves-communications-for-u-s-forces-on-the-move/.

————. "PEO Space Systems Fact Sheet." Accessed February 26, 2019.https://www.public.navy.mil/navwar/PEOSpaceSystems/Documents/146_FactSheet_MUOS5_%202016_FINAL_7Oct16.pdf.

————. Navy Space Implementation, DoD Instruction 5400.43. Washington, DC: Office of the Chief of Naval Operations, 2005.

Dolman, Everett. "Space Force Déjà Vu." *Strategic Studies Quarterly* 13, no. 2 (2019): 16–22. https://www.airuniversity.af.edu/Portals/10/SSQ/documents/Volume-13_Issue-2/Dolman.pdf.

Drew, Jerry V. II, "First in Space: The Army's Role in U.S. Space Efforts, 1938–1958."Space Master's thesis, Naval Postgraduate School, 2017. https://pdfs.semanticscholar.org/4f39/4d47720551819978d26fc185ccb1a8cba553.pdf.

Eidenbach, Peter L. "A Brief History of White Sands Proving Ground: 1941–1965." Accessed September 16, 2019. 2–10. https://web.archive.org/web/20101229102456/http://nmsua.edu/tiopete/files/2008/12/wspgcoldbook.pdf.

Eldridge-Nelson, Allison. "Veil of Protection: Operation Paperclip and the Contrasting Fates of Wernher Von Braun and Arthur Rudolph." Master's thesis, Graduate College of Bowling Green State University, 2017. https://etd.ohiolink.edu/!etd.send_file?accession=bgsu1510914308951993&disposition=inline.

Erwin, Sandra. "Air Force Nominee Barret Calls for Assertive U.S. Posture on Space, Says Space force is A 'Key Imperative'." *Space News*. Last modified September 12, 2019. https://spacenews.com/air-force-nominee-barrett-calls-for-assertive-u-s-posture-on-space/.

———. "Army General to Run One of USSPACECOM's Subordinate Commands." *Space News*. Last modified September 1, 2019. https://spacenews.com/army-general-to-run-one-of-usspacecoms-subordinate-commands/.

———. "Army's Imaging Satellite Up and Running, but its Future Is TBD." *Space News*. Last modified February 21, 2018. https://spacenews.com/armys-imaging-satellite-up-and-running-but-its-future-is-tbd/.

———. "Army Secretary: Still Unclear What Portions of the Army Would Move to the Space Force." *Space News*. Last modified October 8, 2018. https://spacenews.com/army-secretary-still-unclear-what-portions-of-the-army-would-move-to-the-space-force/.

———. "Five Things to Know about the Space Command." *Space News*, Last modified October 23, 2019. https://spacenews.com/five-things-to-know-about-u-s-space-command/.

———. "House Armed Services Markup to Bring Space Force Closer to Reality" *Space News*. June 10, 2019. https://spacenews.com/house-armed-services-markup-to-bring-space-force-closer-to-reality/.

———. "Majority of Army's Space Soldiers will not Transfer to the Space Force." *Space News*. Last modified May 15, 2019. https://spacenews.com/majority-of-armys-space-soldiers-will-not-transfer-to-the-space-force/.

———. "Raymond Endorses Trump's Space Force Proposal." *Space News*. Last modified April 3, 2019. https://spacenews.com/raymond-endorses-trumps-space-force-proposal/.

———. "Shanahan Officially Establishes the Space Development Agency." *Space News*. Last modified March 13, 2019. https://spacenews.com/shanahan-officially-establishes-the-space-development-agency/.

———. "U.S. Space Command's Major Components Will Be Based in California and Colorado." Space *News*. June 30, 2019. https://spacenews.com/u-s-space-commands-major-components-will-be-based-in-california-and-colorado/.

Faulkenberry, Matthew E. "Critical Review of the Navy Space Cadre." Space Master's thesis, Naval Postgraduate School, 2014. https://www.hsdl.org/?view&did=762129.

Federici, Gary A. *From the Sea to the Stars: A Chronicle of the U.S. Navy's Space and Space Related Activities, 1944–2009*. Washington, DC: Department of the Navy, 2010. http://www.history.navy.mil/books/ space/FromTheSeaToTheStars-2010ed.pdf.

Foley, Theresa. "Corona Comes In From The Cold." *Air Force Magazine*. Last modified September 1995. *http://www.airforcemag.com/MagazineArchive/Pages/1995/September%201995/0 995corona.aspx.*

Forest, Benjamin D. "An Analysis of Military Use of Commercial Satellite Communications." Master's thesis, Naval Postgraduate School, 2008. https://core.ac.uk/download/pdf/36697715.pdf.

Green, Constance M., and Milton Lomask. *Vanguard: A History*. Washington, DC: Office of Technology Utilization, 1970. https://history.nasa.gov/SP-42-2.pdf.

Gruss, Mike. "Haney: U.S. Partners to Have Indirect Access to Space Fence Data." *Space News*. Last modified November 21, 2014. https://spacenews.com/42619haney-us-partners-to-have-indirect-access-to-space-fence-data/.

Global Security. "HiHo / Hi-Hoe / NOTS-EV-2 Caleb." https://www.globalsecurity.org/space/systems/hiho.htm.

Hall, R. Cargill., and Jacob Neufeld. *The U.S. Air Force in Space: 1945 to the Twenty-first Century: Proceedings*, Air Force Historical Foundation Symposium, Andrews AFB, Maryland, September 21–22, 1995. Honolulu, HI: University Press of the Pacific, 2002. https://media.defense.gov/2010/Oct/01/2001329745/-1/-1/0/AFD-101001-060.pdf.

Harbaugh, Jennifer. "Biography of Wernher Von Braun." NASA. Last modified August 3, 2017. https://www.nasa.gov/centers/marshall/history/vonbraun/bio.html.

Herb, Jeremy. "House Passes Defense Bill That Would Create 'Space Corps'." *CNN Politics*. Last modified July 14, 2017. https://www.cnn.com/2017/07/14/politics/house-passes-defense-bill-space-corps/index.html.

Howell, Elizabeth. "Classified Shuttle Missions." *Spaceflight*. Last modified October 26, 2016. https://www.space.com/34522-secret-shuttle-missions.html.

Hubbs, Mark. "Mission Begins on Redstone Road." *The Eagle*. October 2007. https://www.smdc.army.mil/Portals/38/Documents/Publications/History/Eagle%20articles/SMDCMissionBeginsonRedstoneRoad.pdf?ver=2019-01-11-144934-020.

Insinna, Valerie. "Documents Reveal How the Space Force Would Launch in 90 Days." *Defense News*. Last modified September 16, 2019. https://www.defensenews.com/digital-show-dailies/2019/09/16/documents-reveal-how-the-space-force-would-launch-in-90-days/.

———."Senate Authorizers Approve Space Force but Switch Up Its Organizational Structure." *Defense News* Last modified May 23, 2019. https://www.defensenews.com/space/2019/05/23/senate-authorizers-approve-space-force-but-switch-up-its-organizational-structure/

———. "Top U.S. Air Force Official Is Now on Board With Trump's Space Force Plan." *Defense News*. Last modified September 5, 2018. https://www.defensenews.com/smr/defense-news-conference/2018/09/05/the-top-air-force-official-is-now-onboard-with-trumps-space-force-plan/.

Jacobsen, Anne. *Operation Paperclip: The Secret Intelligence Program that Brought Nazi scientists to America* (New York: Little, Brown and Company, 2014).

Jewish Virtual Library. "World War II: Operation Paperclip." Accessed September 16, 2019. https://www.defense.gov/About/.

Johnson, Billy E. and Lindsey, Martin F. *U.S. Army Small Space Update*. SSC16-III-06 (SMDC-PA No. 6085, June 3, 2016). https://digitalcommons.usu.edu/cgi/viewcontent.cgi?article=3349&context=smallsat.

Johnson, Kaitlyn. "Space Force or Space Corps?" Center for Strategic and International Studies. Last modified June 27, 2019. https://www.csis.org/analysis/space-force-or-space-corps.https://spacenews.com/dod-working-on-space-force-rollout-plan-pending-congressional-approval/.

Joint Force Space Component Command Public Affairs. "Combined Space Operations
    Center Established at Vandenberg AFB." *Air Force Space Command*. July 19,
    2018. https://www.afspc.af.mil/News/Article-Display/Article/1579285/combined-
    space-operations-center-established-at-vandenberg-afb/.

Judson, Jen, Martin, Jeff, and Gould, Joe. "Here's Who Will be the U.S. Space Command
    Deputy." *Defense News*. Last modified August 6,
    2019.https://www.defensenews.com/digital-show-dailies/smd/2019/08/06/heres-
    who-will-be-the-us-space-command-deputy/.

Kennedy, John F. "President Kennedy's Special Message to the Congress on Urgent
    National Needs, May 25, 1961." JFK Library. Accessed March 8, 2019.
    https://www.jfklibrary.org/archives/other-resources/john-f-kennedy-
    speeches/united-states-congress-special-message-19610525.

Lambeth, Benjamin S. *Mastering the Ultimate Higher Ground: Next Steps in the Military
    Uses of Space* (Santa Monica, CA: RAND, 2003).
    https://www.rand.org/content/dam/rand/pubs/monograph_reports/2005/MR1649.p
    df.

Lasby, Clarence G. *Project Paperclip: German Scientists and the Cold War* (New York:
    Atheneum, 1971).

Leonard, Aaron. "The Horrible Secrets of Operation Paperclip: An Interview with Annie
    Jacobsen about Her Stunning Account." History News Networks. Last modified
    2019. https://historynewsnetwork.org/article/155194.

LePage, Andrew J. "Old Reliable: The Story of the Redstone." The Space Settlement
    Enterprise. Last modified May 2, 2011.
    http://www.thespacereview.com/article/1836/1.

Lewin, Sarah. "'Breaking the Chains of Gravity: The Story of Spaceflight Before NASA'
    (2015): Book Excerpt." Space.com. March 14, 2016.
    https://www.space.com/32192-breaking-chains-of-gravity-book-excerpt.html.

Living Moon. "Project Able." Accessed August 12, 2019.
    https://www.thelivingmoon.com/45jack_files/03files/Project_Able.html.

Lockheed Martin. "Space Fence." Date Accessed November 15, 2019.
    https://www.lockheedmartin.com/en-us/products/space-fence.html

Lorrey, Mike. "Op-ed: The Legal Mandate for a U.S. Space Force." *SpaceNews.com*.
    October 26, 2018. https://spacenews.com/op-ed-the-legal-mandate-for-a-u-s-
    space-force/.

McDonald, Robert A., and Moreno, Sharon K. *Raising the Periscope...Grab and Poppy: America's Early ELINT Satellites*. Chantilly, VA: Center for the Study of National Reconnaissance, 2005. https://www.nro.gov/Portals/65/documents/history/csnr/programs/docs/prog-hist-03.pdf.

McDougall, William A. *The Heavens and the Earth: A Political History of the Space Age*. Baltimore, MD: The John Hopkins University Press, 1985. http://asan.space/wp-content/uploads/2018/05/Walter-A.-McDougall-The-Heavens-and-the-Earth_-A-Political-History-of-the-Space-Age-1997-Johns-Hopkins-University-Press.pdf.

Miller, Clint W. "Optimizing the Navy's Investment in Space Professionals." Master's thesis, Naval Postgraduate School, 2011. https://calhoun.nps.edu/bitstream/handle/10945/5513/11Sep_Miller_C.pdf?sequence=1&isAllowed=y.

Moltz, James C. *The Politics of Space Security: Strategic Restraint and the Pursuit of National Interests*, Second Edition (Stanford, CA: Stanford University Press, 2011).

National Academy of Sciences. Division on Engineering and Physical Sciences. *Navy's Needs in Space for Providing Future Capabilities*. Washington, DC. 2005. https://www.nap.edu/read/11299/.

New World Encyclopedia. "Wernher Von Braun." Last Modified October 19, 2016. https://www.newworldencyclopedia.org/entry/Wernher_von_Braun.

NHCC. "Navy Astronautics Group Established, 22 May 1962," Naval History (blog). U.S. Naval Institute, May 22, 2010. https://www.navalhistory.org/2010/05/22/navy-astronautics-group-established-22-may-1962.

O'Hanlon, Michael. "The Space Force is a Misguided Idea. Congress Should Turn It Down." Brookings Institution. Last modified April 20, 2019. https://www.brookings.edu/blog/order-from-chaos/2019/04/20/the-space-force-is-a-misguided-idea-congress-should-turn-it-down/

Outzen, James. *Critical to U.S. Security: A Compendium of Gambit and Hexagon Satellite Reconnaissance Systems Documents* (Chantilly, VA: Center for the Study of National Reconnaissance, 2012). https://www.nro.gov/Portals/65/documents/history/csnr/gambhex/Docs/Critical%20to%20US%20Security.pdf.

Pawlyk, Oriana. "Former SecAF: Airmen Will Get Lost In 'Space Force'." *DoD Buzz*. Last modified July 30, 2018. https://www.military.com/dodbuzz/2018/07/30/former-secaf-airmen-will-get-lost-space-force.html.

Peebles, Curtis. *High Frontier: The United States Air Force and the Military Space Program*. Washington, DC: Air Force History and Museums Program. 1997. https://media.defense.gov/2010/Dec/02/2001329901/-1/-1/0/AFD-101202-013.pdf

Pence, Mike. "VP Mike Pence Speech on SPACE FORCE Pentagon August 9, 2018." August 9, 2018. Produced by Alberto Lopez. YouTube. 25:47:00. https://www.youtube.com/watch?v=X6maYkfZ514.

RAND Corporation. "A Brief History of Rand." Accessed August 9, 2019. https://www.rand.org/about/history/a-brief-history-of-rand.html.

Richardson, John. "SPAWAR to NAVWAR." June 2, 2019, produced by Naval Information Warfare Systems Command. 2:00:00. https://www.youtube.com/watch?v=y1EJ-3har-I.

Roeder, Tom. "Space Force: A Timeline." *The Gazette*. Last modified June 25, 2018. https://www.coloradopolitics.com/news/space-force-a-timeline/article_307d061b-687c-5332-8286-9b5556c5be61.html.

Rogers, Michael S. "MUOS Satellite Improves Communications for U.S. Forces on the Move." Navy Live (blog). U.S. Navy, July 18, 2013. https://navylive.dodlive.mil/2013/07/18/muos-satellite-improves-communications-for-u-s-forces-on-the-move/.

Rogers, Mike. "Remarks of Congressman Mike Rogers." *Strategic Studies Quarterly 11*, no. 2 (2017): 3–12. https://www.airuniversity.af.edu/Portals/10/SSQ/documents/Volume-11_Issue-2/Rogers.pdf.

Rosenberg, Barry. "Why Is SPAWAR Now NAVWAR? Networks & Cyber Warfare." Breaking Defense, June 5, 2019. https://breakingdefense.com/2019/06/why-is-spawar-now-navwar-networks-cyber-warfare/.

Royston, Ken. "The Army Space Cadre Formal." *Army Space Journal* (2004, no. 1). https://apps.dtic.mil/dtic/tr/fulltext/u2/a523261.pdf.

Sebestyen, Victor. *1946: The Making of the Modern World* (New York: Knopf Doubleday Publishing Group, Nov 29, 2016).

Serena, Katie. "How Nazi Scientist Wernher Von Braun Sent the U.S. to The Moon." ATI. Last modified December 18, 2017. https://allthatsinteresting.com/wernher-von-braun.

SMDC/ARSTRAT Public Affairs. "SMDC Celebrates 60 Years of Space and Missile Defense Excellence." U.S. Army. Last modified September 21, 2017. https://www.army.mil/article/194216/smdc_celebrates_60_years_of_space_and_missile_defense_excellence.

Smith, Marcia S. *Military Space Activities: Highlights of the Rumsfeld Commission Report and Key Organization and Management Issues*. CRS Report No. RS20824. Washington, DC: Congressional Research Service, 2001. *https*://www.everycrsreport.com/files/20010221_RS20824_f0489c442dba0986a6 d7e3ca843938b52559bc9d.pdf

———. "Space Command Gets to Work While Congress Continues to Debate Space Force." *Space Policy Online.* September 2019. https://spacepolicyonline.com/news/space-command-gets-to-work-while-congress-continues-to-debate-space-force/.

———. "Text of Space Policy Directive-4 (SPD-4): Establishing a U.S. Space Force." Space Policy Online. February 19, 2019. (Presidential Memorandum. White House, 2019) .

Smithsonian National Air and space Museum. "Discoverer/ Corona: First U.S. Reconnaissance Satellite." Accessed August 14, 2019, https://airandspace.si.edu/exhibitions/space-race/online/sec400/sec420.htm.

———. "Military Origins of the Space Race." Accessed May 12, 2019. https://airandspace.si.edu/exhibitions/space-race/online/sec200/sec250.htm.

Space Policy Online. February 19, 2019.Presidential Memorandum. White House, 2019 https://spacepolicyonline.com/news/text-of-space-policy-directive-4-spd-4-establishing-a-u-s-space-force/

Spires, David N. *Beyond Horizons: A Half Century of Air Force Space Leadership* (Maxwell AFB, AL: Air University Press, 1998). https://apps.dtic.mil/dtic/tr/fulltext/u2/a355572.pdf

Spires, David N. and Sturdevant, Rick W. *Beyond the Ionosphere: Fifty Years of Satellite Communication* (Washington, DC.: The NASA Series, 1997). https://history.nasa.gov/SP-4217/ch7.htm.

Strout, Nathan. "What Will the Space Development Agency Really Do?"*C4ISRNET*. Last modified July 24, 2019. https://www.c4isrnet.com/battlefield-tech/space/2019/07/24/what-will-the-space-development-agency-really-do/.

Swarts, Phillip. "Air Force Lays Out Its Case for Keeping Space Operations." *Space News*. Last modified May 19, 2017. https://spacenews.com/air-force-lays-out-its-case-for-keeping-space-operations/.

———. "Air Force Sec. Wilson Makes New Space Leadership Position Official." *Space News*. Last modified June 16, 2017. https://spacenews.com/air-force-sec-wilson-makes-new-space-leadership-position-official/.

Trudeau, Arthur C. *Proposal to Establish a Lunar Outpost* (U.S. Army: Chief of Research and Development, 1959). https://history.army.mil/faq/horizon/Horizon_V1.pdf.

Unravel Travel TV. "Huntsville, Alabama "Rocket City, USA" A Spirit of Exploration." March 23, 2019.2:17:00. https://www.youtube.com/watch?v=KsKsU1VeexM.

U.S. Air Force. "Cyberspace as a Domain in which the Air Force Flies and Fights." Last modified November 2, 2006. https://www.af.mil/About-Us/Speeches Archive/Display/Article/143968/cyberspace-as-a-domain-in-which-the-air-force-flies-and-fights/.

———. "Satellite Systems." Accessed August 12, 2019. https://www.losangeles.af.mil/Portals/16/documents/AFD-060912-025.pdf?ver=2016-05-02-112847-777.

U.S. Army. "AMCOM History." U.S. Army. Last modified December 4, 2014. https://www.army.mil/article/139391/amcom_history.

U.S. Army Space and Missile Command. "Future Warfare Center." Accessed Jun 1, 2019. https://www.smdc.army.mil/ORGANIZATION/FWC/.

USASMDC/ARSTRAT. "Army Space Support Teams," U.S. Army, last modified March 23, 2017, https://www.army.mil/standto/2017-03-23.

United States History. "Ronald Reagan`s Military Buildup." https://www.u-s-history.com/pages/h1957.html.

U.S. Navy. "Navy Establishes Naval Network Warfare Command." Last modified March 28, 2002. https://www.navy.mil/submit/display.asp?story_id=1156.

USSTRATCOM. "U.S. Strategic Command Fact Sheet Combined Space Operations Center /614th Air Operations Center." July 2018. https://www.stratcom.mil/Portals/8/Documents/CSpOC_Factsheet_2018.pdf.

Vandenberg Air Force Base. "Evolved Expendable Launch Vehicle (EELV)." Last modified August 4, 2017. https://www.vandenberg.af.mil/About-Us/Fact-Sheets/Display/Article/1266632/evolved-expendable-launch-vehicle-eelv/.

Virts, Terry. "I Was an Astronaut. We Need a Space Force." *The Washington Post.* August 23, 2018. https://www.washingtonpost.com/opinions/i-was-an-astronaut-we-need-a-space-force/2018/08/23/637667e6-a6fb-11e8-b76b-d513a40042f6_story.html?utm_term=.a6d767e3e16e.

Walker, James, Bernstein, Lewis, and Long, Sharon. *Seize the High Ground: The Army in Space and Missile Defense* (U.S. Army Space and Missile defense Command, 2003). https://history.army.mil/html/books/070/70-88-1/cmhPub_70-88-1.pdf.

Wall, Mike. "Alan Shepard's Space Race: Soviet Victory Frustrated First American in Space." Space.com. Last modified May 05, 2011. https://www.space.com/11578-nasa-alan-shepard-space-race-human-spaceflight.html.

———."X-37B Military Space Plane Breaks Record on Latest Mystery Mission." Space.com. Last modified August 26, 2019. https://www.space.com/x-37b-military-space-plane-otv5-duration-record.html.

Weeden, Brian. "Space Force Is More Important Than Space Command," *Texas National Security Review.* Last modified July 8, 2019. https://warontherocks.com/2019/07/space-force-is-more-important-than-space-command/.

Werner, Ben. "Shanahan: Space Force Won't Take Over Navy, Army Space Assets." *USNI News.* Last modified March 20, 2019. https://news.usni.org/2019/03/20/new-space-force-will-not-take-away-navy-space-assets.

Werner, Debra. "Lockheed Martin Prepares to Turn on U.S. Air Force Space Fence on Kwajalein Atoll." *Space News.* Last modified May 3, 2018. https://spacenews.com/lockheed-martin-prepares-to-turn-on-u-s-air-force-space-fence-on-kwajalein-atoll/.

White House. *Remarks by President Trump at a Meeting with the National Space Council and   Signing of Space Policy Directive-3.* Washington, DC: White House, 2018.   https://www.whitehouse.gov/briefings-statements/remarks-president-trump-meeting-   national-space-council-signing-space-policy-directive-3/.

Whittington, Michael. "A Separate Space Force, An 80 year old argument." Maxwell Paper 20, no. 1 (May 2000): 1–19. Air War College.

Williams, Lauren C. "Now's Not the Time for a Space Force, STRATCOM Leader Says." The Business of Federal Technology. Last modified March 21, 2018. https://fcw.com/articles/2018/03/21/space-stratcom-sasc-hyten.aspx.

Wilman, Catherene J. *Space Division: A Chronology, 1980–1984.* Office of History Headquarters. 1991. https://books.google.com/books?id=mXEgedvuspMC&printsec=frontcover&source=gbs_ge_summary_r&cad=0#v=onepage&q&f=false.

Wright, Bruce, "Fighting and Winning in Space- Today and Tomorrow." Air Force Association. Last modified August 14, 2019. https://www.afa.org/publications-news/media/president-perspective.

INITIAL DISTRIBUTION LIST

1. Defense Technical Information Center
    Ft. Belvoir, Virginia

2. Dudley Knox Library
    Naval Postgraduate School
    Monterey, California